T3-BNT-689

MAJOR APPLIANCES
Installing, Troubleshooting, and Servicing

Prentice Hall Career and Technology
Englewood Cliffs, New Jersey 07632

95

#28846837

Library of Congress Cataloging-in-Publication Data

Johnson, Jim.
Major appliances : installing, troubleshooting, and servicing / Jim Johnson
    p. cm.
    Includes index.
    ISBN 0-13-808874-8
    1. Household appliances, Electric—Maintenance and repair.   2. Gas appliances—
Maintenance and repair.   3. Refrigeration and refrigerating machinery—
Maintenance and repair.   I. Title.
TK7018.J65 1994   93-35543
683'.88—dc20   CIP

First, this book is dedicated to Peggy Lee, who has been with me through all the ups, downs, and twists and turns of my professional and personal life for 25 years.

Second, this book is dedicated to the memory of Chuck Johnson. His career in the appliance business began at a time when "servicemen" ran calls for a buck-and-a-half, and one of the fundamental skills necessary was the ability to adjust the latch on a refrigerator door; to a time when "technicians" had to learn to troubleshoot the printed circuit board on a microwave oven.

Acquisitions Editor: Ed Francis
Production Editor: Fred Dahl
Copy Editor: Fred Dahl
Designer: Fred Dahl
Cover Design: Marianne Frasco

© 1994 Prentice Hall Career and Technology
Prentice-Hall, Inc.
A Paramount Communications Company
Englewood Cliffs, New Jersey 07632

Printed in the United States of America
10 9 8 7 6 5 4 3 2 1

ISBN 0-13-808874-8

Prentice-Hall International (UK) Limited, *London*
Prentice-Hall of Australia Pty. Limited, *Sydney*
Prentice-Hall Canada Inc., *Toronto*
Prentice-Hall Hispanoamericana, S.A. *Mexico*
Prentice-Hall of India Private Limited, *New Delhi*
Prentice-Hall of Japan, Inc., *Tokyo*
Prentice-Hall of Southeast Asia Pte. Ltd., *Singapore*
Editora Prentice-Hall do Brasil, Ltda., *Rio de Janeiro*

Contents

List of Illustrations

Preface

There has been a need for a text like *Major Appliances: Installing, Troubleshooting, and Servicing* ever since the business outgrew its infancy and progressed into an industry requiring thousands of trained technicians. Now, due to technological advances in the past decade, coupled with the regulations on refrigerant reclaim, the need for such a text has intensified. The Association of Home Appliance Manufacturers (AHAM) and the National Appliance Retail Dealers Association (NARDA) are predicting a serious shortage of qualified service technicians. The key word is *qualified*. As in any trade, there will be people who will be "working at it," but their performance will be below industry standards.

In addition, many cities and states already have or are considering a certification process for major appliance service technicians. Local or state certifications such as these would be apart from the certifications required by the EPA in regard to the handling of refrigerants. Responsible associations and regulatory agencies consider certification a necessity to ensure industry growth and to prevent serious problems that lead to customer dissatisfaction. Focusing on the challenges facing the appliance service industry, the text offers an overview of the business through a "practical, applied approach, to provide the learner with the necessary information to properly service and install major appliances.

The philosophy behind the text is based on the idea that one cannot be an effective service technician if basic concepts are not fully understood. Everyone understands that electricity makes a refrigerator or washing machine work. But unless you understand where electricity comes from and how it does its job, you cannot follow a schematic diagram and effectively troubleshoot an electrical circuit. If, as a service technician, you have not reconciled that fundamental laws of thermodynamics govern the operation of a refrigeration system, your understanding of the heat transfer concept will be forever shrouded in mystery.

Vocational instructors have long recognized that eliminating the mysteries behind basic concepts is the foundation of technical education and that a confident, able technician can only emerge after this has been accomplished. For this reason, the text provides complete information in regard to the basics of electricity and refrigeration before attempting to apply these principles to the operating functions of refrigerators, washing machines and clothes dryers, dishwashers, gas and electric ranges, and microwave ovens.

ACKNOWLEDGMENTS

I would like to express my appreciation to the many students I've worked with at ABC Technical & Trade Schools and Pima College in appliance classes. Their reaction to ideas and methods of presenting material shaped the focus of this book. I would also like to thank WCI Corporation, Maycor Corporation, Amana Refrigeration Inc., Robinair, TIF Instruments, Gemline Products and Simpson Electric for their contributions in the form of service manuals, catalogs and photographs. Many thanks to Ed Francis for his assistance in wrapping up this project and to Karen Brown for help in getting it started.

Jim Johnson

CHAPTER ONE

Electrical Fundamentals

LEARNING OBJECTIVES   **After studying this Chapter, you will be able to:**

1. Describe the process through which a generating plant converts a raw material into electrical energy.
2. Relate the process of generating electricity from magnetism.
3. Understand why some materials allow electricity to flow easily and why other materials are fair or poor conductors of electricity.

■ ■ ■

Electricity. Ask the average person to explain electricity, and you are likely to hear a brief description of a generating plant and a network of wires leading to homes and offices. For those not employed as a technician in the major appliance service field, a general explanation of an electrical generating station and distribution system would be sufficient. To function as a service technician, however, you'll need a complete understanding of how electrical energy is generated, how it is transported to the desired location, and how it is made to perform useful work.

As a service technician, you will be required to troubleshoot electrical circuits and diagnose system malfunctions through the use of schematic diagrams. To perform these tasks effectively, you must be able to reconcile any questions that may exist in the back of your mind, such as "How is electricity really produced?" or "Why does copper conduct electricity but rubber does not allow electrical energy to flow?"

Once you have answered questions like these and have eliminated any doubts about electricity, its fundamentals, and its relationship to the operation of major appliances, you will be able to function confidently as a service technician. You will not necessarily need to review all the principles discussed in this section every time you're called to solve an electrical problem. But, after studying and understanding the material that follows and filing it away mentally, you will eliminate any mysteries that would otherwise muddle the thought processes required for logical, step-by-step analysis of a malfunctioning piece of equipment.

1.1 PRODUCING ELECTRICITY

Electrical energy begins at a generating station and, in simple terms, is defined as a form of energy that performs useful work when converted to light, heat, or mechanical energy. This definition should be easy for us to accept because we see it in action every day when we turn on a light, use a toaster, operate an automatic washer, cook with an electric range or allow electricity to make our lives easier in many different ways.

The first question is, "How is it produced?" It's common knowledge that the electricity we use to operate appliances comes from a power plant and that a generator is used to produce the energy, but the answer involves a bit more than saying, "It comes from a generating station." It will help you to view the electrical generating station as nothing more than a factory that takes a raw material, such as coal or oil, and changes the chemical energy in these materials into a another form of energy: electricity.

Figure 1-1 illustrates, in its simplest form, the process of converting chemical energy into electrical energy. You'll note that, as you trace the process from its beginning (a mine supplying coal, for example) to its end, the energy takes six different forms.

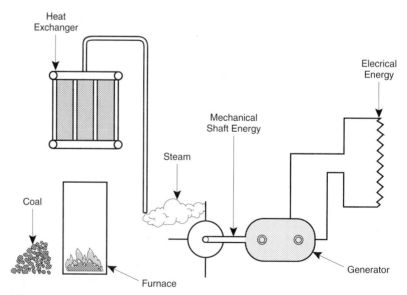

Figure 1-1. Converting chemical energy into heat energy, which creates steam. The high-pressure steam spins the turbine and mechanical shaft energy turns the generator.

In our example, this power plant uses coal as its raw material. Machinery is used to grind the coal into a fine powder, the consistency of which is comparable to face powder. The fine powder is then transported by introducing air pressure into a piping system that leads to furnaces. The pulverized coal is ignited and burned. At this point, the chemical energy is transformed into heat energy.

The heat energy is then used to raise the temperature of water in a closed-loop piping system that is routed through the furnace. The temperature of the water is raised until it boils, changing into steam. The steam is then heated further until its temperature reaches 1,000° Fahrenheit.

Allowing this high-pressure steam to expand and be released in a controlled manner and contact the blades of a turbine, such as the type shown in Figure 1-2, results in a mechanical energy. The mechanical energy is then transferred to the shaft of the turbine, which is connected to the generator rotor. The final result of the energy transformation process is the rotation of the generator that produces electrical energy.

Some electrical generating stations may differ from this illustration in that they use a different source of heat, such as nuclear fission or water power, to achieve mechanical shaft energy, but the final objective is the same: to spin the generators that produce electrical energy.

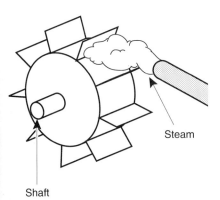

Figure 1-2. High-pressure steam contacting the blade of a turbine results in mechanical energy that spins the shaft connected to a generator.

1.2 ELECTRICITY AND MAGNETISM

The next question is, "How is electricity *really* produced?"

The answer is that the process of generating electricity relates to magnetism. Although magnetism is something we can't see, it's easy for us to accept as being real because we've experienced it when using a bar magnet to pick up pieces of metal. We also understand and accept the fact that there is a magnetic field surrounding the earth. This is proven to us when we use a compass and the needle always points north because of the pull of the earth's magnetic field. The idea of the north and south poles of a magnet setting up a positive/negative field of energy is fundamental to the theory of generating the type of energy used to operate major appliances: alternating current. (We'll discuss alternating current in more detail in Chapter Two.)

The relationship between electricity and magnetism is best explained with the following statement: Cutting the lines of force of a magnetic field with a conductor, such as copper wire, causes a current to be induced in the conductor. This process, discovered in the 1800s and known as *electromagnetism*, is illustrated in Figures 1-3 and 1-4. Figure 1-3 shows a rotor similar to that found in the generator of a power plant. You'll notice that the magnetic energy field sets up from the north pole on one side of the rotor to the south pole on

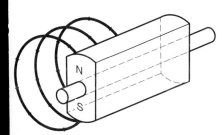

Figure 1-3. Electromagnetic section of a generator. When magnetic lines of force are cut by a conductor, a current is induced in the conductor.

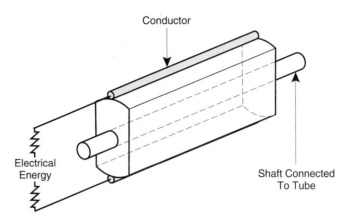

Figure 1-4. A conductor is positioned so that it cuts the lines of force in the magnetic field when the rotor of the generator spins.

the opposite side. Figure 1-4 then, shows a simplified version of the generator itself, with a single conductor positioned in proximity to the generator rotor, allowing the wire to pass through the magnetic lines of force as the rotor spins.

1.3   ATOMIC THEORY

Once electrical energy has been generated, it must be transported along electrical lines before it can perform work. To fully understand how electrical energy is transported, you need to know a little about atomic theory.

All materials, such as metal, glass, or wood, are made up of atoms. An *atom* is described as being the smallest part of an element that can exist alone. Atoms are electrical in structure and are made up of three principal parts: the electron, the neutron, and the proton. Figure 1-5 illustrates an elementary atom. You'll note that the *electron* is shown as carrying a negative charge, and the *proton* is shown to hold a positive charge. The *neutron* has no charge and is said to be "neutral." The neutron and proton combine to form the nucleus of the atom, and the negatively charged electron orbits around the nucleus.

A basic law of physics, known as the *law of electric charges*, states that opposite charges attract and like charges repel. Protons and electrons are attracted to each other because they are of opposite charges. When an atom contains an equal number of electrons and protons, it is said to be *electrically neutral* because there is a balance between the positive and negative charges.

A form of energy induced in a material however, can cause an atom to lose or gain and electron. When this occurs, the energy (electricity) can be transferred through the material, such as a copper wire, until it reaches its destination and performs useful work. Heat

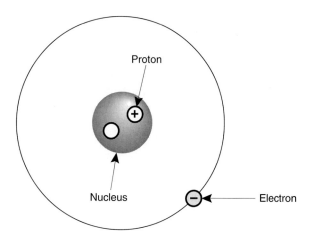

Figure 1-5. An elementary atom. The neutron and proton combine to form the nucleus, around which the electron is in orbit.

energy produced by an electric range bake element and mechanical energy used to run the motor on a dishwasher are two examples of useful work being performed by electrical energy.

The *law of centrifugal force* must also be considered when you're studying atomic theory. This law states that, as an object spins, it will pull away from its center and, with an increase in speed, there is an increase in the force pulling away from the center.

These two laws, the law of electric charges and the law of centrifugal force, along with the atomic makeup of different materials, determine whether or not a substance will be a good conductor of electricity or a poor conductor of electricity.

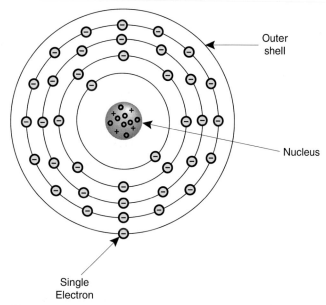

Figure 1-6. A silver atom. Because of its atomic structure, silver is an excellent conductor of electrical energy.

1.4   CONDUCTORS VS. INSULATORS

The atomic structure of a material determines whether or not it will allow electrical energy to flow easily. Silver, for example, is one of the best conductors of electricity because in its atomic structure, there are five orbits or shells of electrons that surround the nucleus of the atom. Figure 1-6 illustrates a silver atom. As you can see in the illustration, the outermost electron is identified as the *valence electron*. Identifying this single electron and understanding that it is spinning at the highest velocity possible without actually breaking free from the force that holds it in position (according to the law of electric charges) is the key to understanding how electrical energy travels through a conductor.

Undisturbed by an energy force, the electron will remain in the orbit because the negatively charged electron is attracted to the positively charged proton in the nucleus of the atom. When a form of energy, such as electricity, is introduced, however, this valence electron is knocked out of its orbit, as shown in Figure 1-7.

In practical terms, this electron is "knocked out of orbit" because of the distance between the free electron and the nucleus. Distance from the nucleus determines the speed of the electron in its orbit. The greater the distance, the faster the electron spins.

Once this process of causing a valence electron to leave its orbit begins, it continues from atom to atom until the energy reaches the end of the conductor. Imagine a piece of wire 5 feet in length. If you were to cause energy to be induced into one end of the wire by cutting the lines of force of a magnetic field, this energy would flow through the wire at a rate that we would, for all practical purposes, consider instantaneous.

Let's review a few key phrases in the preceding paragraph. First, we used the phrase "cause energy to be induced" in reference to generating electricity. The key word here is *induced*, which is another form of the word *induction*, a common term used to describe the method of generating electricity. Second, we said that "energy would flow" once the process of generating electricity is initiated. This flow of energy can also be referred to correctly as *electron flow*, since the movement of the energy actually takes place because of the transferring of an electron from one atom to another.

To ensure your firm understanding of electron flow in a conductor, we'll compare the ability of two more conductors—copper and aluminum—to conduct electrical energy. The copper atom shown in Figure 1-8 demonstrates that the atomic structure of copper differs from that of silver. You'll recall that a silver atom contains five orbits, or shells, of electrons. The copper atom contains only four orbits. Because of this difference, silver is a better conductor of electricity. The easiest approach to understanding this concept is to recall the law of centrifugal force. We said that the speed at which an electron spins around the nucleus increases as the distance from the nucleus

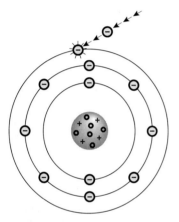

Figure 1-7. The single valence electron of an atom of a conductor is easily knocked out of its orbit by electrical energy.

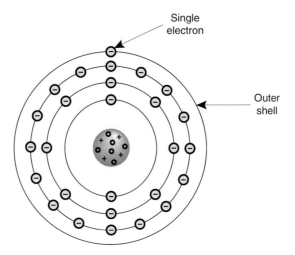

Figure 1-8. A copper atom. Copper is not the best of all conductors. Because of its atomic structure, it has less energy than silver.

increases. This increased speed and higher electron count causes the atom to contain a higher level of energy. The higher the level of energy there is, the better the electron flow.

This principle also explains why copper is a better conductor than aluminum. The atomic structure of aluminum does not allow electrons to flow as easily as copper. Therefore, copper does a better job of transporting electrical energy from its source to its destination.

Although silver is the best conductor of electricity, it is not widely used due to its high cost. In major appliances, silver is used on contact points in some components such as refrigerator thermostats and electric range surface unit switches. We'll be discussing these components in detail in later units.

In the same way that the atomic structure of a material can allow electrons to flow easily, it can also make electron flow difficult.

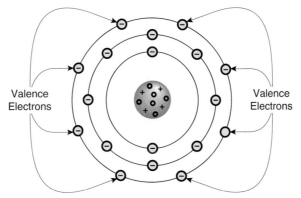

Figure 1-9. The atomic structure of an insulating material differs from that of a conductive material. While conductor atoms have only one valence electron that is easily knocked out of orbit, the atom of an insulator has up to 8 valence electrons.

Figure 1-9 illustrates a material that would be classed as an *insulator* rather than a conductor. You'll note that, instead of only one valence electron spinning around the nucleus as in a conductor, there are eight electrons in the outer shell of an insulator atom. This difference in atomic structure is what enables you to hold an insulated wire, such as a refrigerator cord connected to a "hot" circuit, in your hand without experiencing an electrical shock.

In the case of a conductor, the striking electron gives up all of its energy to one electron. But in an insulator, the energy carried by a single valence electron is divided among all eight electrons. As a result, the electrical energy cannot travel along the insulator material because it is, in effect, dissipated, having little effect on the atom. In other words, the insulating material is in contact with the surface of the conducting material, but the different atomic makeup of the material retards the flow of energy.

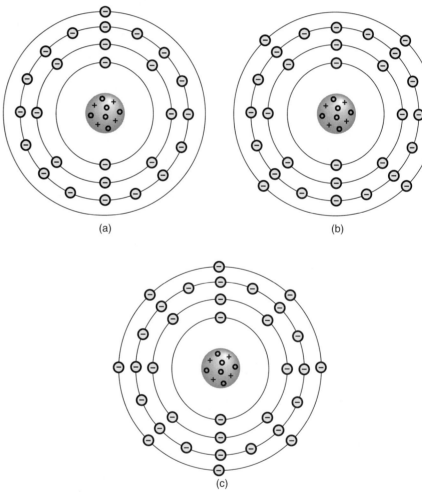

Figure 1-10. Conductors, semiconductors, and insulators differ in their atomic structure. A conductor atom (a) has only one valence electron, a semiconductor atom, (b) will have up to 4 valence electrons and an insulator atom, (c) contains up to 8 valence electrons.

While conductors are said to contain free electrons, it is said of insulators that they contain bound electrons, or electrons that are not easily knocked out of their orbit. These electrons will remain bound, however, only as long as the level of energy induced is not too high. There is no perfect insulator and, under the right conditions, such as extremely high voltage being transported along the wires connecting the power plant to homes and offices, an insulator will "break down" and allow electron flow.

Some example of insulators are the rubber or plastic that surrounds the wire used to conduct electricity to various components in major appliances. Wood and glass are also insulators and poor conductors of electricity.

1.5   SEMICONDUCTORS

Although we will be covering semiconductors used in solid state components used in appliances in detail in a later chapter, a brief explanation of semiconductor material fits in here since the basic atomic theory applies to the conductivity of all materials.

As the name implies, a semiconductor is neither a good conductor of electricity nor an effective insulator. Figure 1-10 shows a comparison between conductors, semiconductors, and insulators. While the conductor atom material contains only one valence electron and an insulator atom has eight valence electrons, the semiconductor atom is classified as being at midpoint between the two extremes, having four valence electrons. In reality, conductors may have one or two valence electrons, semiconductors may have three or four valence electrons, and insulators may have seven or eight valence electrons.

The chart in Figure 1-11 will help you to understand the difference in conductivity of various materials. As you can see, the materials are rated according to their resistance to electron flow, with conductors near the bottom of the scale and insulators near the top of the scale.

In the manufacture of semiconductors, particularly in solid-state components, two materials known as germanium and silicon, are the most popularly used. You've probably heard the term *silicon* used before in reference to an area in California known as the "Silicon Valley." This area was so named because of the large concentration of electronic manufacturing firms located there. In the last decade, the use of solid state components, such as printed circuit boards, in major appliances has increased dramatically.

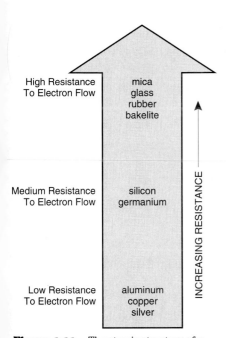

Figure 1-11. The atomic structure of a material dictates whether it will be a good conductor of electrical energy or if it will be an effective insulator.

CHAPTER ONE   SUMMARY

Electrical energy is a mystery to the average person. But, to function as an appliance service technician, you must have a complete understanding how electricity is generated and how it travels along electrical lines and through circuitry in major appliances.

Electrical energy begins at what is fundamentally a "factory" that uses a raw material, such as coal or oil, to create mechanical energy to turn a generator. Some electrical generating stations may be nuclear or hydroelectric.

The electrical energy used to operate major appliances is alternating current, known as *electromagnetic energy*. Electromagnetic energy is induced in a conductor, such as a copper wire, when the conductor passes through or "cuts the lines of force" of a magnetic field. This process was discovered in the 1800s.

Electrical energy can be conducted through some materials easily, while others resist the flow of electrons. The atomic makeup of a material determines whether or not it will be a conductor or an insulator. There is no perfect insulator, and if the level of energy is high enough, an insulator will break down and allow electrical energy to flow. Silver, copper, and aluminum are good conductors of electricity, while materials such as rubber, plastic, glass, or wood are classed as insulators.

A semiconductor is a material that is neither a good conductor of electricity nor an effective insulator. Semiconductor materials are used in the manufacture of solid state devices used in printed circuit boards and other solid state controls used in major appliances.

Alternating Current Fundamentals, Terms and Definitions

LEARNING OBJECTIVES

After studying this Chapter, you will be able to:

1. Define the method through which alternating current is generated.
2. Define the terms *volt, ampere, resistance,* and *watt.*
3. Understand how Ohm's law is used to solve electrical problems and illustrate the relationship between the three basic electrical units: current, voltage, and resistance.
4. Understand how the cost of electrical power is calculated and use a step-by-step method to calculate the cost of operating an appliance.

■ ■ ■

2.1 ALTERNATING CURRENT

When working as an appliance service technician, you should develop the habit of looking for the manufacturer's equipment information tag as you begin servicing the refrigerator, washing machine, or whatever it is you happen to be repairing. Experienced techs consult the tag for the model and serial numbers (especially if the appliance is under warranty) and look for the operating data of the item. One piece of information you'll encounter on the model tag relates specifically to the type of energy used to power the appliance.

Terms such as *120 VAC* or *240 VAC* tell you the proper level and type of energy that must be applied for the appliance to operate properly. *VAC* is an abbreviation for *volts alternating current,* and the numbers preceding the abbreviation tell you the proper voltage necessary for various pieces of equipment.

A refrigerator, for example, operates through a power cord that is plugged into a standard wall outlet, the same receptacle into which you could plug a table lamp or an electric drill. A standard outlet of this type delivers 120 VAC to the item when it is plugged in. An electric range or standard size electric dryer, however, requires a higher level of energy to operate. When consulting the equipment tag on items such as these, you'll find that 240 VAC is required. You'll also notice a big difference in the type of power cord (usually referred to as a *pigtail*) used on the higher voltage appliances.

The source of all alternating current is the AC generator. As you'll recall from Chapter One, we said that electrical energy is generated when a conductor is passed through the lines of force of a magnetic field. It is more correct to say that the conductor is *rotated* as it passes through the magnetic field. The problem with this description is that, for some, it brings to mind an image of a rotating wire—and then a question: "If the wire is rotating, why doesn't it get twisted?" This question is raised because it's common knowledge that the conductor is fastened tight to a connection that allows electrical energy to flow along the distribution system. The initial reaction to this description of a conductor rotating just doesn't fit with this knowledge.

In practice, one way to apply the rotating conductor definition is to provide some type of slip ring connection that would allow contact of the rotating portion of the conductor to the stationary portion. Another more commonly used method is to let the conductor remain stationary and attach the magnets to the rotating portion of the generator. Figure 2-1 illustrates this concept which eliminates the need for a slip ring method. Cutting the lines of force of the magnetic field is still accomplished when the magnets are connected to a shaft that is rotated through mechanical means.

Once you have accepted the fact that electrical energy is generated through the electromagnetic principle, understanding alternating current is easy. All alternating current generated in the United States is delivered to homes and offices at a rate (or frequency) of 60 cycles

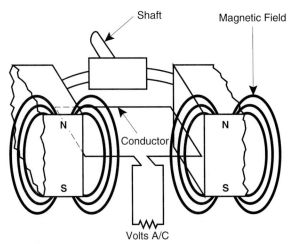

Figure 2-1. An elementary AC generator in which the conductor is fixed and the magnets are allowed to spin around the conductor.

per second. In many instances you'll encounter this information listed on manufacturers equipment tags as *60 HZ*. The abbreviation *HZ* stands for *hertz*, a term meaning cycles per second. (The term is derived from the name of a German physicist, Heinrich Rudolph Hertz, who proved during the 1800s that electromagnetic waves existed.)

This rate, or frequency, of energy delivery is accomplished through the rotation of the generator at the proper RPM (revolutions per minute). To achieve 60 hertz, the 2-pole generator must rotate at 3,600 RPM. This figure becomes logical through some simple multiplication. Multiplying 60 cycles per second by 60 (60 seconds in each minute) yields a result of 3,600. Simply stated then, a conductor cutting the lines of force of a magnet rotating at 3,600 RPM will deliver electrical energy at a frequency of 60 hertz.

To have a complete understanding of alternating current, you should also have a firm understanding of two more terms: sine wave and polarity.

Polarity refers to the poles of a magnet used in a generator. Each magnet has a north pole and a south pole. As the magnet rotates, the energy generated changes in polarity from negative to positive.

This concept of energy changing from a negative peak to a positive peak is explained in Figure 2-2, which shows the sine wave. You'll note that the peak values of the sine wave, both negative and positive, are shown. One complete rotation of the generator is represented by the line that begins at the intersecting point of the horizontal and vertical lines, then rises to the positive peak value before crossing the horizontal line, drops to the negative peak value point, and returns to the starting point on the horizontal line. The time frame shown is 1/60th of a second. Sixty complete rotations of the magnet in 1 second are necessary to deliver alternating current at a frequency of 60 hertz.

As we said, this explanation applies to generating stations in the United States. While operating principles are the same in European

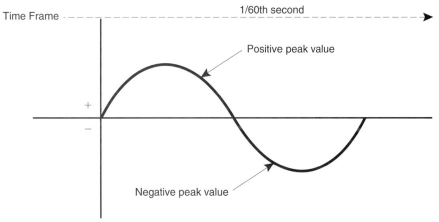

Figure 2-2.   The sine wave illustrates one 360° rotation of an AC generator.

countries, the result is different due to a slower rotation of the magnetic field. In other countries, alternating current is delivered at a frequency of 50 hertz.

This information is relevant because, as a technician, you may encounter an appliance that requires conversion from one system to the other if your customer takes the item from one country to another. The Amana corporation, for example, has manufactured microwave ovens in the past that can operate on either a 50- or 60-hertz system after conversion by a qualified service technician.

Caution! *When servicing an appliance in this manner always consult the manufacturer's service manual for specific instructions in regard to the particular model of appliance.*

2.2   VOLTAGE

Voltage is a term representing the level of electrical potential in a circuit. Sometimes referred to as *electromotive force*, the concept of voltage is easiest to understand when related to the idea of pressure. Electrical energy is generated in a conductor, and the voltage is the level of pressure that "pushes" the energy along the conductor. While the concept of electrical pressure may not be easy for you to relate to, this idea will be simpler for you to understand when it is compared to pressure in a water line.

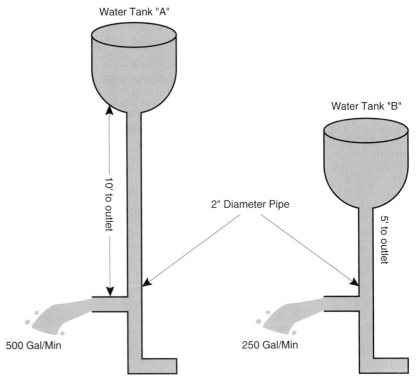

Figure 2-3.   Because of the higher elevation of tank A, a greater level of pressure is produced at the outlet although the same diameter pipe is used for both tanks.

Figure 2-3 illustrates this idea by showing the concept of varying pressure. You'll note that the pressure at the outlet of tank A is shown to be 500 gallons per minute and that tank B is delivering at a pressure of 250 gallons per minute. The difference in pressure is due to the difference in elevation between the two tanks.

This concept applies to an electrical circuit. Simply compare the longer length of pipe and higher tank to a generator of a larger capacity that is capable of delivering electrical energy at a pressure greater than that of a generator of a smaller capacity.

When the voltage in an electrical circuit is increased, the current flow is also increased. You could illustrate this concept by connecting a 120-volt light bulb to a 240-volt circuit. The light bulb would burn much brighter when connected to the circuit of higher pressure. Of course, the bulb wouldn't last very long since it isn't designed to operate at the higher voltage. Applying 240 volts to a motor designed to operate on 120 volts will cause it to burn out. Conversely, applying 120 volts to an electric oven bake element designed to operate on 240 volts will result in a very low heat output by the element.

The term *volt* is used in honor of an Italian professor of physics, Alessandro Volta, who discovered in 1800 that electricity was produced by a chemical action between moisture and two dissimilar metals.

2.3 CURRENT/AMPERES

Current is defined as the rate at which electrons flow through a conductor. When measuring the number of electrons that flow past a given point of a conductor, you use the term *ampere*. This term is derived from the name of Andre Ampere, a French physicist who, in 1820, was able to measure the magnetic effect of electrical current.

Technicians working with major appliances approach this concept from the point of view that they must be concerned with the *amp* (or *amperage*) *draw* of a given electrical component. All electrical components that perform work have a given amperage draw (correctly referred to as *current draw*) when they are energized. If the correct voltage is applied to a component and it is not defective, it will do its job and draw the correct amount of current. If an incorrect voltage is applied or if it is defective, the amperage draw will be incorrect.

You can troubleshoot a defrost circuit on a refrigerator, for example, by determining that the radiant heater is drawing the correct amperage and is therefore providing heat as it is designed to do. Or you could diagnose a compressor as being defective if you understand this principle and are able to apply it in the practical application of testing an electrical component.

Conductors are correctly sized and insulated according to the amperage draw of the component to which they will supply energy. This concept will be easier to understand if you relate it to a thin extension cord that is supplying power to more items than it is designed to handle. The cord will overheat because the combined current draw of all the items connected to it requires that too much current flow through the undersized conductor.

From a service standpoint, one of the fundamental things to keep in mind about amperes is that, if the voltage is too low, then the amperage draw will be too high. As an example, let's say that, because of a bad connection, a refrigerator compressor that is designed to draw 2.5 amps when the correct voltage (120 VAC) is applied, is only receiving 80 percent of the voltage required for proper operation. In the event that this excessive voltage drop occurred, the compressor's amperage draw would be higher than normal.

The expression "amp draw" (as opposed to "amperage draw") is used, in many cases, in manufacturer's service manuals and on equipment tags. In some cases, you may even find the letter *A* on an equipment tag to indicate that a particular component, such as a motor or a heating element, draws a given amount of current. The refrigerator compressor we discussed previously, for example, could be listed as having a current draw of 2.5 A or 2.5 amps.

The rate of current flow in an electrical circuit is measured with an *ammeter*, a type of meter we'll be discussing in detail in a later chapter.

2.4 RESISTANCE

The opposition to current flow (the flow of electrons along a conductor) is referred to as *resistance*. There is some resistance in all conductors. Even if the atomic makeup of a material allows electrical energy to flow and we refer to the material as a "good conductor of electricity," there is still some resistance to the flow of electrons. An electrical component in an appliance, such as a heating element or a motor winding, will also have a given amount of resistance.

This degree of resistance is measured in ohms. When you encounter information about the level of resistance of a component on a manufacturer's wiring diagram, the term *ohm* is represented by the Greek symbol Ω (omega). The term is derived from the name of a German teacher, Georg Ohm, who made discoveries about the relationship between the electrical quantities of resistance, voltage, and amps in 1826. The level of resistance of a given conductor or electrical component is measured with an *ohmmeter*.

Resistance to electron flow is the cause of voltage drop and heat in a conductor. As we said in our discussion on amperes, an excessive amount of heat can be created in a conductor if it is too small to carry the current load required to operate several different items. The ability of resistance to create heat, though, is also used in some major appliances to do work. Some conductive materials, such as those used in electric range elements or the heating elements used in electric clothes dryers, are designed to have a very high resistance to current flow and, as a result, they offer a great amount of heat.

2.5 OHM'S LAW

Ohm's law is an important concept for an appliance service technician to understand because it illustrates the relationship between the three basic electrical units: current (amps), voltage, and resistance.

Ohm's law involves the use of letters to represent the electrical units. The letter E represents voltage, because it stands for electromotive force. The letter I represents current or amps flowing in a circuit, because it stands for induction. (You'll remember from Chapter One that we said it was correct to say that a current is *induced* in a circuit.) The letter R represents resistance in an electrical circuit.

These three letters are used in three formulas that allow you to calculate the unknown factor of a circuit (voltage, current, or resistance), providing you know or can measure two of the factors.

$$E = I \times R$$

Multiplying the amperage (I) by the resistance (R) will tell you the voltage (E) applied in a circuit.

$$I = E \div R$$

Dividing the voltage (E) by the resistance (R) will tell you the amps (I) flowing in a circuit.

$$R = E \div I$$

Dividing the voltage (E) by the amperes (I) will tell you the resistance (R) in a circuit.

Unless you've been exposed to formulas in a chemistry class or a similar setting, the idea of using this method where the factor you want to find is listed first can take some getting used to. But after you practice with the concept—$E = I \times R$ or voltage is equal to amps times the resistance)—you find it fairly easy to use.

Other than getting used to formulas, you may be comfortable with a memory aid such as the one shown in Figure 2-4. You'll note that this method uses a memory circle, with the electrical units shown. Covering the appropriate letter allows you to use the formula that offers the solution to the problem.

Figure 2-5 illustrates a circuit that offers you an opportunity to apply Ohm's law. The circuit shows a bake element in an electric range. The heating element has a resistance of 24 ohms and the voltage applied to the circuit is 240 VAC. The memory circle can be used to determine the amperage draw (current flow) in the circuit. Covering the letter I (for current, the factor we want to find) shows that the voltage must be divided by the resistance.

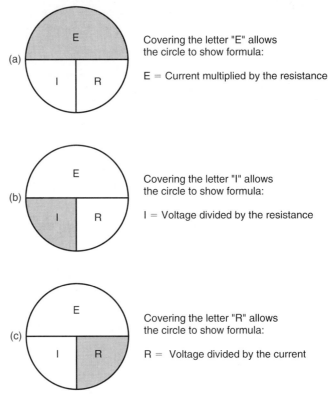

(a) Covering the letter "E" allows the circle to show formula:

E = Current multiplied by the resistance

(b) Covering the letter "I" allows the circle to show formula:

I = Voltage divided by the resistance

(c) Covering the letter "R" allows the circle to show formula:

R = Voltage divided by the current

Figure 2-4. The ohm's law memory circle can be used to simplify the formulas used to illustrate the relationship between voltage, current, and resistance.

Of course, this illustration makes use of even numbers and ideal conditions, something you'll rarely encounter in field applications. As the bake element heats up, for example, the resistance in the element changes. This would cause the calculated current to differ from the actual ammeter reading. A slight fluctuation in the voltage, either up or down, would also cause meter readings that would not compare exactly to the calculations.

2.6   ELECTRICAL POWER/WATTS

We said at the beginning of Chapter One that electricity is simply defined as a form of energy that performs useful work when converted to heat, light, or mechanical energy. When electrical energy is converted to another form of energy, the energy is then referred to as *electrical power*, and this power is measured in watts. A brief definition of *watt* is the unit of measurement of true power.

The term *watt* is probably much more familiar to you than ohm ampere, and volt are, largely because you've been exposed to it in your everyday activities. You've used a hair dryer that is rated in watts, and you're used to the idea that the wattage rating means that you'll get a certain amount of work (heat, in this case) out of the dryer. You've

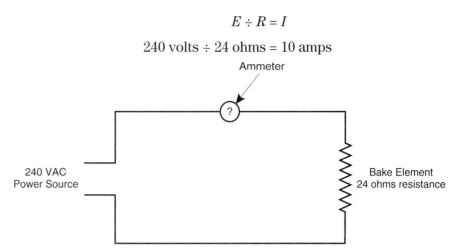

$$E \div R = I$$

$$240 \text{ volts} \div 24 \text{ ohms} = 10 \text{ amps}$$

Figure 2-5. The memory circle can be used to calculate the current in this circuit and determine what the reading on the ammeter should be. If any two values are known, the third value can be determined through the use of Ohm's law.

changed a burned-out light bulb, replacing it with a new one that matches the old one in wattage rating. And it's common knowledge to us that a 100-watt bulb will put out more light than a 60-watt bulb.

Just as a 100-watt light bulb will be brighter than a 60-watt bulb, an appliance technician must understand that this concept applies to electrical components, such as motors and heating elements used in refrigerators or ranges. A motor that is designed to pull a heavy load, such as a drive motor on an automatic washer will have a higher wattage rating than a motor that is designed to provide air flow in a refrigerator, and a small heating element will have a lower wattage rating than a larger element.

When working as an appliance technician, you must also understand the relationship between watts and amps. In some cases, manufacturers will list information on an equipment tag and use the term *watt*, while on the others will use the term *amp* in regard to the operating characteristics of a refrigerator or microwave oven.

From a practical standpoint, both terms relate to the same concept—(the amount of electrical energy used)—and understanding the relationship between the two is a simple task.

If, for example, a refrigerator manufacturer listed information about the compressor of the unit in terms of amps when referring to the current draw of the component, the technician could calculate the power in watts by means of a simple formula:

$$P = E \times I$$

In simple terms, this means P (power in watts) is equal to E (electromotive force in volts) multiplied by I (current in amperes). Or, if you're the type of person who is uncomfortable with this method that seemingly lists things backwards, you could say it this way: Volts times amps equals watts.

As an example, use the bake element we referred to in Figure 2-5 when explaining Ohm's law. If you worked the exercise in the illustration, you found out that the amperage draw of the element was 10 amps. If you proceed a step further and use the preceding formula, you can calculate the watts (that is, measure the true power of the bake element circuit):

$$240 \text{ volts} \times 10 \text{ amps} = 2,400 \text{ watts}$$

You've used the two known quantities, volts and amps, to calculate the wattage of the element.

This information is relevant because it relates to the cost of operating the electric range. The utility company that supplies electrical energy measures the rate of power consumption in kilowatts (KW). One *kilowatt* is equal to 1,000 watts, and charges for energy used are listed in kilowatt-hours (KW-H).

The utility company, then, uses still another formula to determine the actual dollar and cents cost of operating the element. The formula reads as follows:

$$E = \frac{P \times T}{1,000}$$

E (energy, in this case, not electromotive force as in the other formulas) is equal to P (power in watts) multiplied by T (time), which is then divided by 1,000 (a figure representative of a kilowatt).

To determine the annual operating cost of the element, we need (in addition to the manufacturer's amp rating and the applied voltage) the following factors:

1. The number of hours per day the element is used.
2. The charge by the electric company per kilowatt.

Let's look at an example to tie it all together.

Amp draw of the element: 10 amps

Applied voltage: 240 VAC

Hours or use per day: 2

Charge per kilowatt hour: 9 cents

And, to determine the yearly operating cost of the element, we could use the following steps:

STEP 1: Determine the power consumption in watts.

$$P = E \times I$$
$$2,400 \text{ watts} = 240 \text{ VAC} \times 10 \text{ amps}$$

STEP 2: Determine the power consumption in kilowatt-hours for one 24-hour period.

$$E = \frac{P \times T}{1,000}$$

$$\frac{2,400 \text{ watts} \times 2 \text{ hours}}{1,000} = 4.8 \text{ kilowatts/hours}$$

STEP 3: Determine the operating cost in a 24-hour period at 9 cents per kilowatt-hour and multiply by 365 days in the calendar year.

4.8 kilowatt-hours $\times$ 0.09 cents per kilowatt $\times$ 365 days = $157.68

Our conclusion, then, is that in a situation where a bake element that is rated at 10 amps in current draw, where the on-time is 2 hours per day, and where the electric company charges 9 cents per kilowatt-hour, the yearly cost of operating the element is $157.68.

One thing you have to keep in mind about this exercise is that we have been working with an electric range bake element, which is a purely *resistive load*. If we were presented with the problem of calculating the yearly operating cost of a different kind of load, such as a refrigerator compressor, we would have to use a slightly different method. The reason is that a refrigerator compressor is an electric motor and an electric motor is known as an *inductive load*.

Appliance manufacturers use these same procedures to determine the consumer information they are required to provide in regard to the estimated operating cost of their products. Upon a visit to an appliance store, you would find the estimated annual operating cost posted on each product.

CHAPTER TWO SUMMARY

The type of energy supplied to major appliances is alternating current. Refrigerators cperate on 120 VAC (volts alternating current). Other appliances, such as electric ranges and standard size electric dryers, require higher voltage to operate properly (240 VAC).

The source for alternating current is the AC generator, which is rotated at a speed of 3,600 RPM in order to achieve an energy delivery at a rate of 60 hertz (60 cycles per second). To produce alternating current, the lines of force of a magnet are cut by a conductor, and this negative/positive energy field is best understood when the terms *polarity* and *sine wave* are understood. Alternating current is generated at a frequency of 60 hertz in the United States, but may be generated at 50 hertz in other countries.

Voltage relates to the electrical pressure in a circuit and is sometimes referred to as *electromotive force*. All electrical components are designed to operate on a given voltage and perform a given amount of work (light, heat, or mechanical energy) when the correct voltage is applied. Applying excessive voltage to a component will cause it to burn out. Applying less voltage than required will yield less than the proper amount of work.

Current flow in a circuit is measured in *amperes*, the rate at which electrons flow past a given point in a circuit. When an electrical circuit is properly designed and constructed, the wire is sized properly and the correct insulation is used so that the conductor can carry the proper amount of current. Overcurrent running through a conductor that is too small can cause heat or fire, or result in damage to the equipment.

Resistance is measured in *ohms*. All conductors offer some resistance to current flow in a circuit. Some electrical components, such as heating elements for electric ranges or clothes dryers, are designed to present a very high resistance to current flow. The end result is that heat is generated.

Ohm's law is used to illustrate the relationship between volts, current, and the resistance of a circuit. If two factors are known, the formulas can be used to calculate the third unknown factor.

The unit of measurement of true power in a circuit is the watt. There is a relationship between the terms *watts* and *amps*, and the wattage draw of a component can be calculated by multiplying the volts times the amps in a circuit. The utility company measures the amount of energy used in kilowatts and charges at a set fee per kilowatt-hour. One *kilowatt* is equal to 1,000 watts. When dealing with a purely resistive load, such as a bake element in an oven, calculating the cost of operation is done by means of a simple formula. In the case of an inductive load, such as an electric motor, a different formula must be used.

CHAPTER THREE

Electrical Distribution Systems

LEARNING OBJECTIVES   **After studying this Chapter, you will be able to:**
1. Understand why the electric company generates power at an extremely high level of energy when the level of power required by the consumer is much lower.
2. Describe the method of construction of a step-up and step-down transformer.
3. Identify the four basic current/voltage systems supplied to customers by the electric company.
4. Understand how an electrical distribution system within a residence delivers power to various types of equipment.

■ ■ ■

3.1 ELECTRIC COMPANY POWER DISTRIBUTION

Although we've discussed the theory behind the generation of electrical energy, you can't see the electricity. It would be easy to harbor some doubt as to whether or not the things we've talked about are, in fact, real. Despite any doubts about the generation of electrical energy because it can't be seen, we all accept the fact that electricity is something that can be felt, especially when we consider the high-voltage lines that connect our homes to the utility that supplies us with power. To grasp how the alternating current is transported from its point of origin to our homes, where it performs the useful work of operating major appliances, you need only recall the concepts discussed in prior chapters.

Figure 3-1 shows a typical generator you might see on a visit to an electrical generating plant, most of which have several such generators. This one is approximately 5 feet high at its highest point, 8 feet

Figure 3-1. A typical generator found in an electric plant. Most electrical generating stations use several such generators to supply power to a service area.

wide, and 15 feet long. As you can imagine, a generator of this size, containing large magnets capable of a very strong magnetic field and accommodating many conductors, is capable of delivering a great amount of energy.

In addition to using generators of large capacity, the electric company also uses another component, the transformer, to achieve the high level of energy (in many cases from 120,00 to 220,000 volts) that is necessary to service a surrounding community. A *transformer* is a device that receives a certain voltage at its primary winding and, depending on whether it is a step-up or step-down transformer, delivers either a higher or a lower level of energy from its secondary winding.

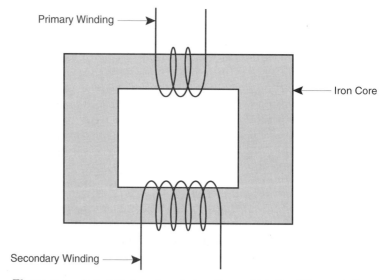

Figure 3-2. A simplified version of a transformer. Note the difference in the number of turns of wire in each winding.

Figure 3-2 shows the fundamental design of a transformer. It is a simple device that is easily understood. The transformer is actually an iron core around which wire of a given number of turns of wire is wound on one side (the *primary winding*), and a different number of turns of wire is wound around the opposite side (the *secondary winding*).

You'll note that the transformer shown has more turns of wire in the secondary winding than in the primary winding. This is because we're illustrating a basic *step-up transformer*. In the case of a *step-down transformer*, the secondary winding would have fewer turns than the primary winding. A transformer used at the electric company is a step-up transformer, which boosts the voltage generated to a high level.

To better understand how a transformer works, we'll go a bit deeper into its method of construction from a practical point of view. If you hold a small transformer in your hand and look closely at the windings that are wrapped around the iron core, you'll notice that, although the windings are copper in appearance, the wire itself is, in fact, insulated. A very thin, yet strong, insulation is wrapped around the conductor. There is no metal-to-metal contact between the winding and the laminated iron core. Given this fact, the question arises, "If the bare wire is not touching the iron core of the transformer, how does the electrical energy get from one winding to another?"

To answer that question, refer to Chapter One, in which we discussed electricity and magnetism. You'll recall that we said electrical energy is generated in a conductor when that conductor cuts the lines of force of a magnetic field. What you need to recognize about this phenomenon is that it is directly reversible. In other words, just as a current flow is set up in a conductor by passing the conductor through a magnetic field, a magnetic field is set up around a conductor that has current flowing through it. The magnetic field surrounding the conductor sets up lines of force that are carried through the iron core of the transformer. The secondary winding, then, since it is also wound around the iron core, cuts the lines of force passing through the core and picks up the energy (voltage). This is how the transformer receives a primary voltage, and, through the secondary winding which is made up of more turns than the primary winding, steps up the voltage before it is sent along the high-voltage lines that supply energy to the end user.

Another common question is often raised in the discussion of electrical distribution systems: "If appliances operate on relatively low voltages (120 and 240 VAC), why must the electric company step up to such a high level of energy (120,000 to 220,000 VAC)?"

The answer to this question lies in the fundamental concept of atomic theory, which states that all conductors offer some resistance to current flow and as a result voltage drop occurs. To ensure that the proper voltage is delivered to their customers, electric companies boost voltages to an extremely high level of energy to compensate for the drop that takes place within the miles of wiring that connects the generating station to homes, offices, and manufacturing plants.

At the end of the electrical distribution system, there is another transformer. This transformer may be located on a pole outside the customer's residence or on the ground near their homes or commercial buildings. This is a step-down transformer that is designed to accept a high level of energy at its primary winding and deliver a lower level of energy from its secondary winding. This method of the step-down transformer provides the proper voltage to the end user.

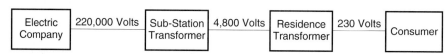

Figure 3-3. The electrical distribution system from the source to the consumer.

Figure 3-3 illustrates a typical electrical distribution system, through which you can trace electrical energy from its point of origin, through a substation, and finally to the consumer. This network of wiring and transformers is referred to as a *grid*.

3.2 VOLTAGE, PHASE, AND CURRENT SYSTEMS

In Chapter Two, we said that, as an appliance service technician, you should develop the habit of looking for the manufacturer's equipment tag to determine operating factors, such as the voltage required and the amperage draw of the unit. When referring to voltage, you may see this on an equipment tag as *240 VAC 1 PH*.

The *PH* stands for *phase*, a term that refers to the number of AC sine wave patterns occurring at different times within an AC circuit. This difference in phase is accomplished by variation of generator design and the distribution of power by the electric company. In our example, the manufacturer has advised that this particular piece of equipment operates on a 240-volt, single-phase system. The voltage we're discussing would most likely be applied to an electric range or a full-sized electric clothes dryer. On some equipment tags, you may see the symbol ∅ to represent the term *phase*.

An electric company commonly provides electrical energy to the consumer at several different voltage levels and with different characteristics of current. The four basic systems of current\voltage are:

240-volt, single-phase.
240-volt, three-phase.
208-volt, three-phase.
480-volt, three-phase.

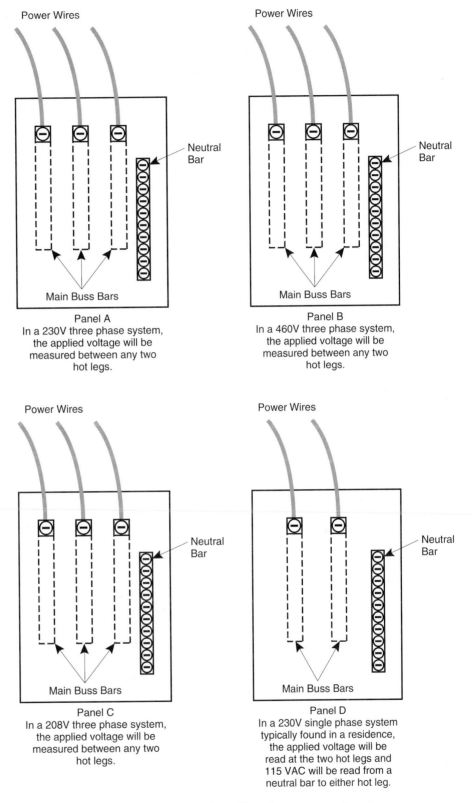

Figure 3-4. The main disconnect panels used in various power supply systems.

A 480-volt, three-phase system is used only in industrial applications and a 240-volt and 208-volt, three-phase system would be found only in light commercial applications. The 240-volt, single-phase system is the standard for residences. Obviously, as an appliance tech, you'll be dealing with residential systems most of the time, but you should have an understanding of the other systems since you may be called on to service an appliance in use in other than a residence.

Suppose, for example, a customer in a light commercial building finds it necessary to connect a standard electric clothes dryer to a 208-volt system. An electrician taps into the main circuit panel and runs a single-phase circuit to the dryer. The customer may wind up complaining that the dryer "takes too long" to dry damp items. The reason for this complaint is not, in fact, a malfunction of the dryer, but is related to the fact that the dryer is designed to operate on 240 VAC. Applying a lower-than-called-for voltage to the heating element, such as that found in a clothes dryer, would result in that element doing less work than it is designed to do. Operating on a lower-than-called-for voltage means that the element wouldn't be putting out as much heat as it is designed to deliver, and this would result in the dryer taking "too long" to dry clothes.

Figure 3-4 offers a practical illustration of the variations in electrical energy supplied to customers by showing the basic layout of the main electrical panel for each system. You'll note that the "hot" wires (*hot legs*, as they are referred to by electricians) are identified. Panel D also explains the method through which 120 volts are supplied to operate refrigerators, dishwashers, and microwave ovens, as well as other 120-volt items.

3.3 ELECTRICAL DISTRIBUTION SYSTEM WITHIN A RESIDENCE

When the 240-volt, single-phase system is delivered from the secondary winding of the residence transformer, the wiring is connected to a main electrical panel that serves as a distribution center for power to the electrical system within a residence. A main disconnect panel may be mounted on an exterior wall of the home in close proximity to the electric meter, or it may be located in a laundry room, furnace room, or basement.

Older homes use a group of screw-in type fuses, known as *plug fuses*, in conjunction with a pull-out cartridge with two fuses that serves as a main disconnect. A second pull-out assembly, with cartridge-type fuses, usually supplies power to the electric range or electric furnace if the house is so equipped.

This combination of cartridge and plug fuses provides overload protection on the various circuits. A circuit breaker that can be reset is now used in modern residences. Figure 3-5 shows a comparison between the older plug fuse panel and the modern circuit breaker

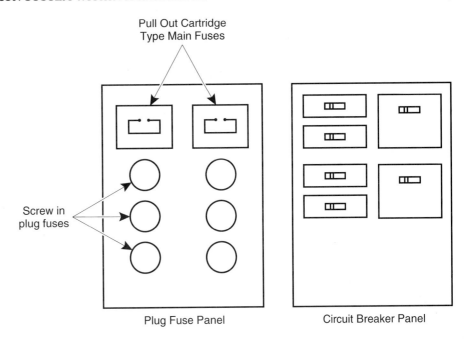

Figure 3-5. A plug fuse panel found in older homes and a circuit breaker panel used in new homes. The circuit breaker panel will contain single-pole and two-pole breakers.

panel. You'll notice that the main panel contains two sizes of circuit breakers: single-pole breakers and two-pole breakers. A *two-pole breaker* will provide a 240-volt circuit to items such as an electric range or full size electric dryer. The *single-pole breakers*, in addition to supplying 120-volt circuits for lighting and wall receptacles, supply a power circuit for refrigerators, dishwashers, automatic washers, and microwave ovens.

As an appliance service technician, you will, at some point trace an equipment problem to the main power distribution panel. Although you will not be required to replace circuit breakers or repair wiring within the panel, you must be able to confirm that the problem is located in the wiring system rather than in the appliance. Why? Because you must keep in mind that your customer only knows one thing: The appliance doesn't work, and it's up to you as the technician to inform the customer of the problem.

A loose connection at a circuit breaker, for example, would present a higher-than-normal resistance at the point and as a result would mean that the voltage delivered to the wall receptacle will be lower than it should be. If a refrigerator is plugged into the wall receptacle, it wouldn't work properly due to the incorrect level of power being applied. This lower-than-normal voltage would result in higher-than-normal amperage draw. Appliances, such as a refrigerator, are equipped with overcurrent protection devices designed to shut the appliance down in the event of higher-than-normal amperage draw.

Circuit breakers snap into the main circuit panel and attach to the main buss bars in several different ways, depending on the manufacturer.

Caution! *If you find it necessary to remove the cover on a main circuit panel, use extreme care! The wires connected to the top of the panel are "hot" because they are the wires leading from the power supply to the building. Turning off circuit breakers does not kill the power at these connections. The circuit breakers, when shut off, only shut down the power to the wires directly connected to them. Carelessness could cause severe electrical shock that will result in serious injury or death!*

In a typical wiring system for a residence, the electrical energy is applied to the main disconnect panel. Conductors leading from the panel to wall receptacles, lighting systems, electric water heaters, and furnaces, or to wall outlets for electric dryers and electric ranges deliver power throughout the network of wiring.

Heavier wiring is used for the electric range and dryer outlets, electric furnaces, and electric water heaters. This is due to the fact that these items are classified as high-amperage equipment. They require 240 VAC for their proper operation, while the power delivered to the standard wall receptacles is run through lighter-gauge wire. The 120-volt system is accomplished by providing one hot leg to the wall outlet, then connecting a neutral wire that provides a complete circuit.

CHAPTER THREE   SUMMARY

To supply the amount of energy required by home, office, and manufacturing plants in an area, an electrical generating station utilizes a number of generators capable of delivering a large amount of power. In addition to generators, the utility company uses step-up transformers to boost the energy to an extremely high level (120,000 to 220,000 VAC). This is necessary because all conductors offer some resistance to current flow and voltage drop occurs as electrical energy is being transported over a long-distance wiring system. A *transformer* is an electrical component that receives a given level of energy at its primary winding and puts out either a higher or lower level of energy from its secondary winding. At the beginning of the electrical distribution system, a *step-up transformer* is used; at the end of the distribution network, a *step-down transformer* is used.

Depending on the customer, the electric company will provide one of four basic current/voltage systems. An appliance technician will usually be concerned with a 240-VAC, single-phase system. An understanding of other systems is essential in order to troubleshoot problems with appliances used in commercial applications.

The main electrical panel serves as a distribution center for power within a residence. The panel may be located outside or in a basement, laundry room, or furnace room. Older homes used plug and cartridge fuses for circuit protection; more modern residences have circuit breakers in their distribution panels. A blown fuse must be replaced but circuit breakers may be reset. A two-pole breaker provides 240-volt service and a singlepole breaker provides 120-volt service. Two hot legs supply power to a 240-volt circuit, while one hot leg and a neutral wire provide power to a 120-volt circuit.

Caution! *Always use extreme care when working near a main disconnect panel. Carelessness could result in severe electrical shock.*

Electrical Components for Major Appliances

After studying this Chapter, you will be able to:

1. Identify the different types of components used in refrigerators and freezers, laundry equipment, and cooking appliances.
2. Differentiate between different types of heating elements used in various appliances.
3. Identify a component as a switch or a load.

■ ■ ■

Now that you understand where electricity *really* comes from, as well as how it is transported from the point of generation and distributed throughout a residence, it's time to start applying this fundamental knowledge to the components used in major appliances. To benefit most from this unit, stay focused on the basic definition of electricity: a form of energy that performs useful work when converted to light, heat, or mechanical energy.

When it comes to refrigerators, laundry, or cooking equipment, this basic definition applies on all counts. We use mechanical energy (motors) in several ways, and light for customer viewing of the product or as cycle indicators. We apply electrical power to a wide range of heating elements used in cooking equipment, dishwashing equipment, refrigerators, and electric clothes dryers.

4.1 MOTORS

As an appliance service technician, you'll be replacing motors of many different sizes and styles, used in a wide range of applications. Some motors will be small enough to fit in the palm of your hand,

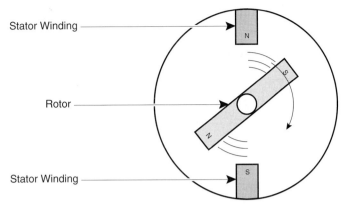

Figure 4-1. An elementary electric motor. The magnetic field set up by the stator winding causes the rotor to spin.

and some will be so heavy and large that it will be a chore to hold them in place while you install them. Regardless of the size, style, or application, the first step in learning to troubleshoot and diagnose motor problems is to understand the elementary electric motor.

Figure 4-1 shows the basic components of an electric motor and illustrates the basic concept (magnetism) behind its operation.

As you can see, the two parts of our elementary motor are the *stator* (the section that *stays* in place) and the *rotor* (the section that *rotates*). Our rotor is really nothing more than a magnet and our stator is the section that contains electrical windings. The law of electrical charges (like charges repel and unlike charges attract) is what makes a motor operate. When electrical energy is applied to the stator windings, a field of energy is set up, specifically an *electromagnetic* field.

Electromagnetic is a key word for you if you fully understood our explanation of the generation of electrical energy in Chapter One. You'll recall we said that, when a conductor is rotated through a magnetic field, a form of energy is set up in the conductor. To understand electric motors, the theory you have to buy into is this: There is a direct correlation between the theory of electromagnetism creating a form of energy in a conductor and the conductor then creating its own magnetic field.

This is illustrated in Figure 4-2. Drawing A shows a conductor positioned in a magnetic field in which rotation cuts the lines of force and sets up a form of energy that can pass along the conductor. Drawing B shows the energy passing along the length of the conductor and, as a result, creating a magnetic field that surrounds the conductor.

To understand how the rotor section of a motor is made to spin, apply the theory of a magnetic field surrounding a conductor in the following manner: The *stator* of the motor is made up of an iron core around which is wound the conductor of a given diameter and length (quite long when you consider the number of turns in a motor winding). This conductor is energized with electricity because it has electrical current flowing through it. A very thin, very tough insulation surrounds the wire and prevents any electrical short that would occur when an

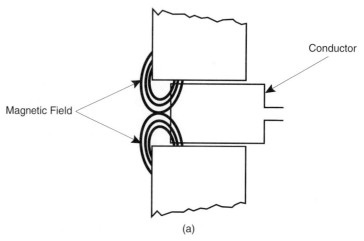

Figure 4-2 (A). When a conductor cuts the lines of force of a conductor, a current flows through the conductor.

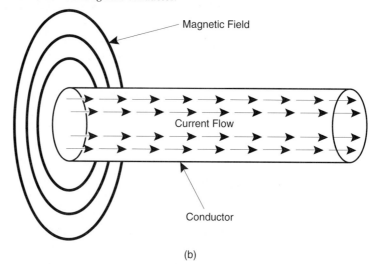

Figure 4-2 (B). As current flows through a conductor, a magnetic field is set up around the conductor.

uninsulated wire touches metal. The method of winding the wire around the stator will determine many of the characteristics of a motor.

Current flow in the stator winding, coupled with the lines of force of the rotor magnet, cause the rotor to spin when the windings are energized. This rotation is actually accomplished by the continuous and sequential repulsion (caused by like charges) and attraction (caused by unlike charges) in the magnets. The fact that we're applying alternating current plays a part in determining the pole of the stator magnet at any given moment. Due to the characteristic of alternating current, the poles of the stator windings change from north to south.

The "push and pull" of the magnets, combined with the centrifugal force that propels the rotor past a point it has been either attracted to or repulsed from, results in a continuous spinning of the rotor.

All that remains is to attach a shaft to the rotor. (This would be done in a manner that prevents the shaft from becoming magnetic.) And, depending on the application in which our motor is used, we would attach either a pulley that would drive a belt, a fan blade that would create air flow, or, in a refrigerator compressor, a crankshaft that causes a piston to move up and down.

Some refrigerator compressors are not piston-types (reciprocating) compressors but instead are referred to as *rotary compressors*. In this case, the rotor is attached to a different type of assembly. (We'll discuss compressor types in more detail in later chapters.)

4.2 MOTOR APPLICATIONS/CAPACITORS

Motors are used to accomplish different tasks in the operation of a major appliance and, as a result, there are many different types of motors. Size (horsepower rating), application, and cost are three factors that determine the design and method of construction of a motor.

In some instances, such as an automatic washer motor, high starting torque is necessary and a significant amount of mechanical energy is required to accomplish the task at hand. An *induction-start/induction-run motor*, which is usually of $\frac{1}{2}$-HP capacity, would be used. A *start capacitor*, which is an electrical component that delivers a momentary boost to give the motor even more starting torque, is used by many manufacturers.

Another type of motor you'll encounter when working as an appliance service technician is the *shaded pole motor*. This motor is much smaller than automatic washer motors and because of different construction methods, has a very low starting and running torque. A shaded pole motor is commonly used to circulate air flow in a frost-free refrigerator.

Window air conditioning units will commonly use *PSC (permanent split capacitor)* motors. In this instance, a *run capacitor*, which is a component similar to a start capacitor, is used to make the motor run more efficiently.

Very small induction-type motors are used to operate ice makers in refrigerators or to operate timing devices in automatic washers, clothes dryers, microwave ovens, and refrigerators. Page 36 shows some of the different types of motors and motor-related components used in appliances.

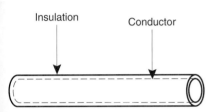

Figure 4-3. The conductor that makes up the windings of a motor is insulated with a very strong insulator.

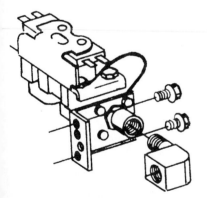

Figure 4-4. A solenoid-operated water valve. The types of mounting brackets and fitting used will vary depending on the application of the valve.

4.3 SOLENOIDS

A solenoid-operated water valve, such as the one shown in Figure 4-4, is used in dishwashers. This type of device also uses an electrically energized coil of wire to create an electromagnet. Unlike a motor,

Washing Machine Motors

Clothes Dryer Motors

Various motors and components
used in major appliances

Dishwasher Motors

Refrigerator Evaporator Fan Motors

Refrigerator Condenser
Cooling Fan Motor

Run Capacitor

Start Capacitor

Motors and capacitors. Various motors and capacitors used in major appliances. *(Courtesy Gem Products, Inc.)*

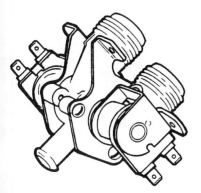

Figure 4-5. A double solenoid valve used in washing machines.

Figure 4-6. A double solenoid water valve used in a refrigerator equipped with an ice maker and chilled water dispenser. This type of valve has a single inlet and two outlets.

however, the electromagnetic field does not spin a rotor, but instead causes a rod to move to a different position, overcoming spring pressure, thereby creating an opening that will allow water to flow.

A single solenoid-operated valve will also be found in a refrigerator equipped with an ice maker. Energizing the solenoid coil for a brief period of time allows a small amount of water to flow into the ice maker assembly.

A *double solenoid valve*, also known as a *mixing valve*, is used in automatic washers. (Refer to Figure 4-5.) In this application, one solenoid may be energized to allow cold water to flow into the automatic washer, one solenoid may be energized to allow hot water only, or both solenoids may be energized to allow a mixture of warm water to enter the automatic washer tub.

A refrigerator equipped with an ice maker and a chilled water feature will also use a double solenoid water valve, but in this case it is not a mixing valve.

Figure 4-6 shows this style of double solenoid valve. You'll note that there are two outlet connections for water flow, as opposed to the mixing valve shown in Figure 4-5, which has two inlets for water but only one outlet.

Other uses for solenoids are for the operation of locking devices on self-cleaning ranges, as door locks on some models of microwave ovens, on dispensers of fabric softener in automatic washers, or as part of an off-balance switch assembly in some makes of automatic washers. Page 39 shows some of the applications of solenoids in appliances.

4.4 HEATING ELEMENTS

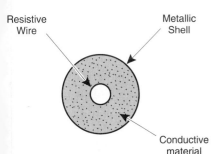

Resistive Wire

Metallic Shell

Conductive material

Figure 4-7. A bake element consists of a resistive conductor surrounded by a heat conductive material wrapped in a metallic outer shell.

A heating element used in appliances is an electrical component that is highly resistive to electron flow and, as a result, produces a high level of heat whenever electrical energy is applied.

One application of heating elements is in electric ranges. Surface units and bake elements are manufactured in a wide variety of styles and wattage ratings. Pages 40–43 show a sampling of various sizes and configurations used by different manufacturers.

Technicians sometimes refer to this type of element as a *Cal-rod type*. Cal-rod is a brand name that has become accepted by many as the technical description for this style of element (just as, for example, many plastic laminate countertop materials are referred to as "Formica" when in fact the material is manufactured by other companies).

If you were to carefully cut through one of these elements with a hacksaw, you would find that it is made up of a thin resistive wire, surrounded by an asbestos-type of powdery, heat-conductive material encased in a metallic shell. (See Figure 4-7.) When an element of this type is energized, the shell will glow red-hot, changing from its normal black color.

Other applications for this type of element are as a water heating element in dishwashers and as a defrost heater in frost-free refrigerators. Page 40 shows some of the many different shapes of heaters used by manufacturers in these applications. While some may be similar to bake elements in color and of a very simple shape, others may be aluminum-colored and of a more complicated shape. Some are designed to wrap around both sides of the tubing assembly it is supposed to keep clear of frost.

Caution! *All heaters, regardless of shape and color, can cause a painful burn if touched while they are energized.*

Another popular type of heating element used in appliances is the ribbon-type element. This component is also a resistive wire, but unlike the Cal-rod type of element, it is not surrounded by a conductive material encased in a metal jacket.

When the ribbon-type element is used in electric dryers or in an assembly energized during the dry cycle of a dishwasher, the element is positioned in a manner that will allow air flow directly over its surface. Ceramic insulators are used to prevent the element from contacting a metal surface. When used as a defrost heater in a frost-free refrigerator, however, the ribbon-type element is encased in a glass tube. A ribbon-type element will glow bright red when energized and operating properly. Page 43 shows some of the various configurations and applications of the ribbon type element.

In addition to the types of heating elements already discussed, you'll encounter several other styles of elements when working with appliances. They range from very low-wattage heaters made of stranded wire wrapped in rubber or plastic insulation, to a metal element that has a braided appearance. You may even see some heaters glued to a heavy foil backing. (See page 44.)

Uses for these elements are as drain trough heaters, as dew point compensators to prevent refrigerator cabinets from sweating, as heaters around openings for dispensing ice, or as an aid to operating an ice maker or to eject cubes. Page 44 shows some examples of these elements that are used in specialized applications. We'll be discussing these components in more detail in later units.

Up to now, our discussion has centered around electrical components in appliances that use electrical energy to perform useful work. A component of this description is referred to as a *load*, in other words, a device to which we apply power. Now we'll discuss components that are not loads, but switches. *Switches* don't use electrical energy; they allow only it to pass onto loads.

We use switches to energize and deenergize a load according to how much heat or cooling we want, or to the length of cycle we desire. Thermostats (sometimes referred to as cold controls when we're talking about refrigerators instead of ovens), motor starting relays, pressure-sensing devices, water temperature selectors, bimetal devices, and timers are all a switch in one form or another.

Ice Maker Water Valves

Dishwasher Water Valve

Washing Machine Mixing Valves

Various solenoid operated
components used in
major appliances

Transmission Shifting Solenoid

Agitate/Spin
Shifting Solenoid

Solenoid-operated components. Used in many different applications in major appliances. *(Courtesy Gem Products, Inc.)*

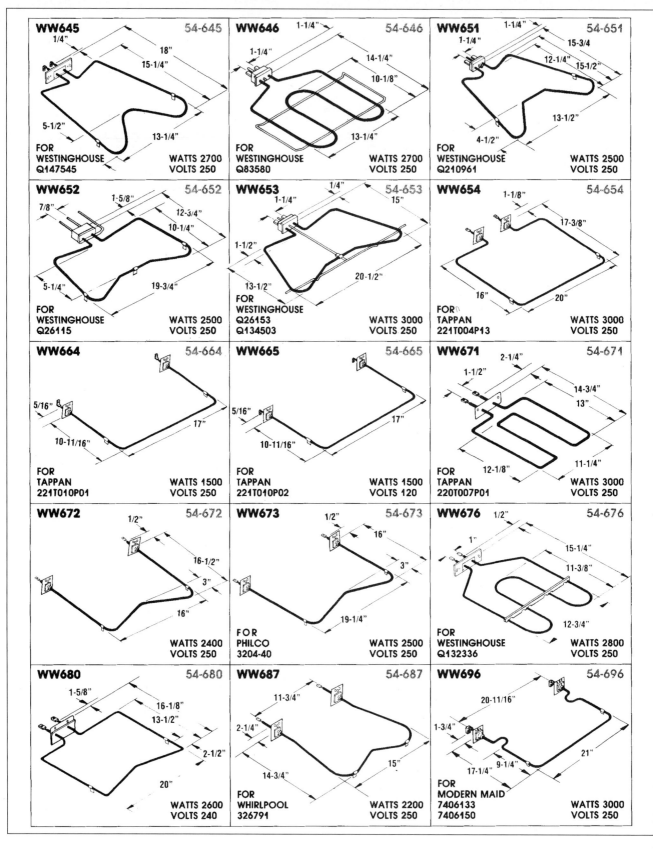

Bake and broil elements. A wide range of bake and broil elements are used in different makes of cooking equipment. (*Courtesy Robertshaw*)

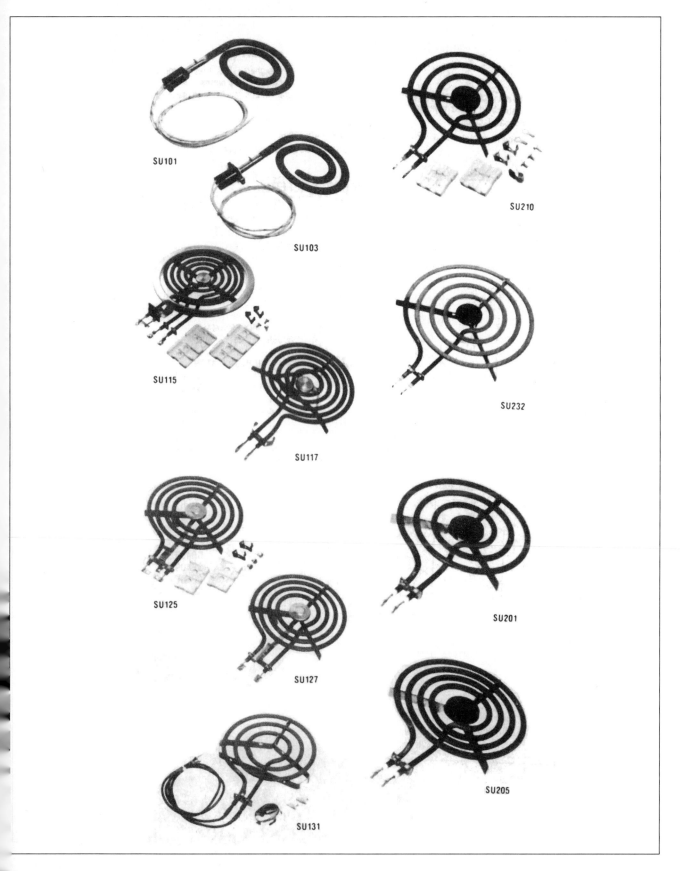

SU101

SU103

SU210

SU115

SU117

SU232

SU125

SU127

SU201

SU131

SU205

Surface unit elements.   Various styles and sizes of surface units used in electric ranges. *(Courtesy Gem Products, Inc.)*

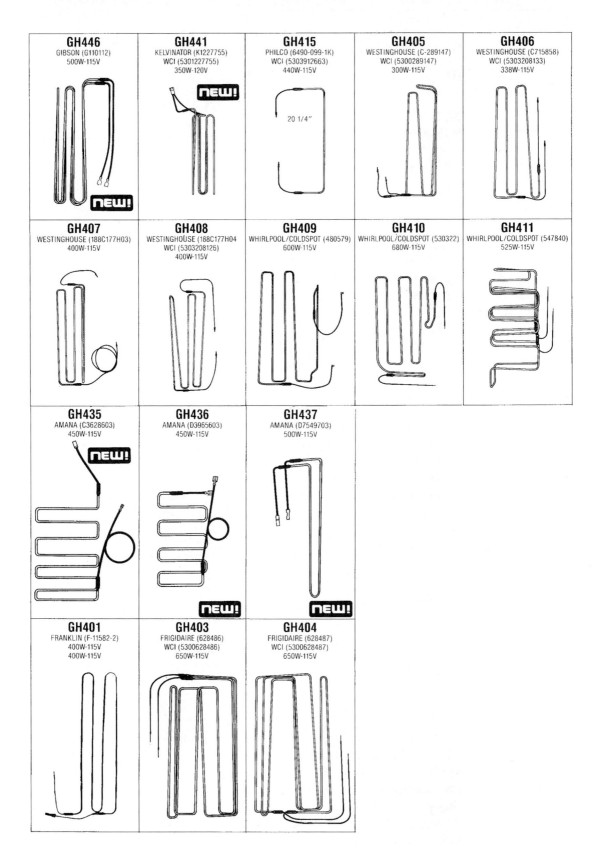

GH446	GH441	GH415	GH405	GH406
GIBSON (G110112) 500W-115V	KELVINATOR (K1227755) WCI (5301227755) 350W-120V	PHILCO (6490-099-1K) WCI (5303912663) 440W-115V	WESTINGHOUSE (C-289147) WCI (5300289147) 300W-115V	WESTINGHOUSE (C715858) WCI (5303208133) 338W-115V

GH407	GH408	GH409	GH410	GH411
WESTINGHOUSE (188C177H03) 400W-115V	WESTINGHOUSE (188C177H04) WCI (5303208126) 400W-115V	WHIRLPOOL/COLDSPOT (480579) 600W-115V	WHIRLPOOL/COLDSPOT (530322) 680W-115V	WHIRLPOOL/COLDSPOT (547840) 525W-115V

GH435	GH436	GH437
AMANA (C3628603) 450W-115V	AMANA (D3965603) 450W-115V	AMANA (D7549703) 500W-115V

GH401	GH403	GH404
FRANKLIN (F-11582-2) 400W-115V 400W-115V	FRIGIDAIRE (628486) WCI (5300628486) 650W-115V	FRIGIDAIRE (628487) WCI (5300628487) 650W-115V

Tubular defrost heaters. Depending on the make and model of refrigerator, many different configurations of defrost heaters are used. *(Courtesy Gem Products, Inc.)*

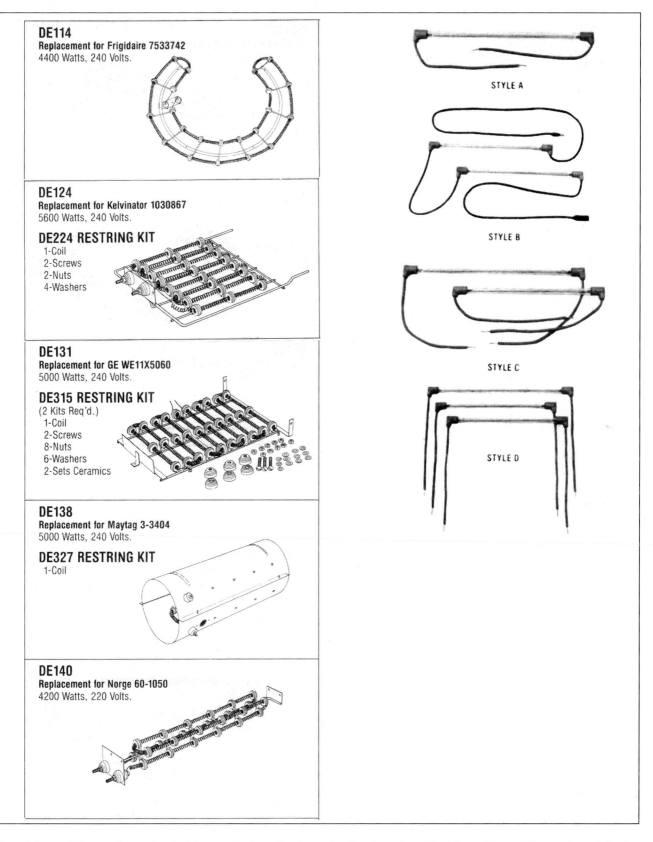

DE114
Replacement for Frigidaire 7533742
4400 Watts, 240 Volts.

DE124
Replacement for Kelvinator 1030867
5600 Watts, 240 Volts.

DE224 RESTRING KIT
1-Coil
2-Screws
2-Nuts
4-Washers

DE131
Replacement for GE WE11X5060
5000 Watts, 240 Volts.

DE315 RESTRING KIT
(2 Kits Req'd.)
1-Coil
2-Screws
8-Nuts
6-Washers
2-Sets Ceramics

DE138
Replacement for Maytag 3-3404
5000 Watts, 240 Volts.

DE327 RESTRING KIT
1-Coil

DE140
Replacement for Norge 60-1050
4200 Watts, 220 Volts.

STYLE A

STYLE B

STYLE C

STYLE D

Glass tube and dryer elements. A nichrome wire is another type of heating element used in major appliances. When used as a defrost heater in a refrigerator, the heating element is contained in a glass tube. Some manufacturers may warn that the technician should not touch the glass portion of the defrost heater during installation. *(Courtesy Gem Products, Inc.)*

43

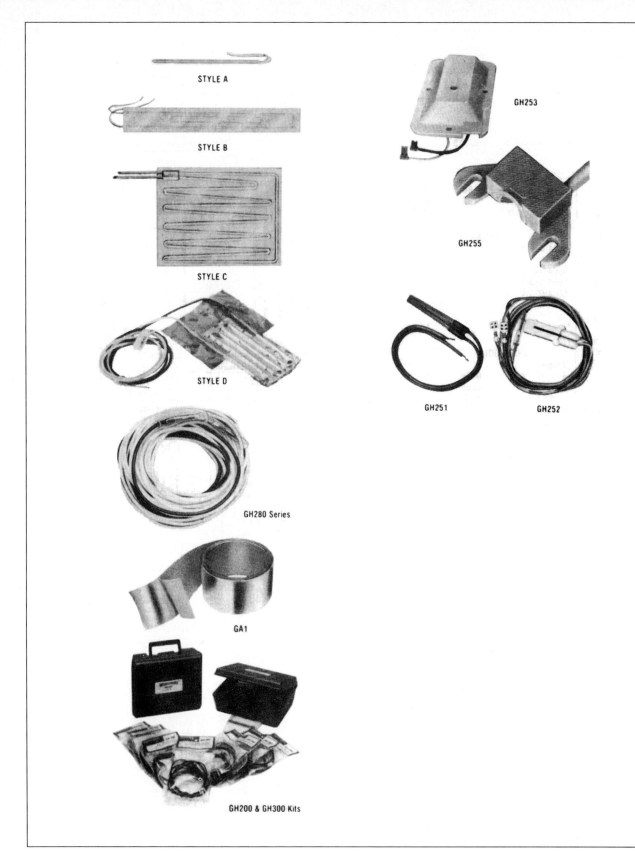

STYLE A

STYLE B

STYLE C

STYLE D

GH253

GH255

GH251   GH252

GH280 Series

GA1

GH200 & GH300 Kits

Low wattage heaters. Low wattage heaters are used in many different applications in major appliances. In some cases the technician may use a kit form of heater to custom fit a replacement. *(Courtesy Gem Products, Inc.)*

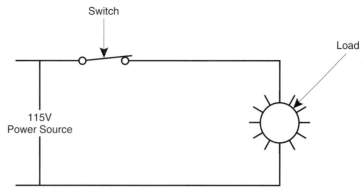

Figure 4-8. A simple electrical circuit. A switch allows current to pass, and a load performs useful work when electrical energy is applied.

And, while they may *seem* to be complex and confusing, the principle behind their use is no more complicated than the application of the simple light switch and bulb circuit shown in Figure 4-8. As you can see from the illustration, a circuit is completed to the light bulb (load) when the switch is in its closed position. When you flip up the switch on the wall, a complete path for current flow is provided and the bulb lights up. Then you flip the switch back down, the circuit is broken and the bulb goes off.

Switches in appliances perform the same basic function. In some cases they may be manual, starting or stopping a fan, or turning a light off and on just like the switch on the wall. But in some cases the switches react to temperature or pressure, or they may be part of a motor-driven device that uses a cam assembly to make and break a circuit. Regardless of the method of operation, they all perform the same function: either allow or prevent current flow to a load.

4.5   THERMOSTATS

It's the thermostat that allows the appliance owner to choose an oven temperature of 400° Fahrenheit, operate their clothes dryer on permanent press rather than high heat, or select (within certain guidelines) the operating temperature of their refrigerator.

As shown in Figure 4-9, the thermostat is a device that reacts to temperature. This particular illustration is related to the operation of a refrigerator. Drawing A shows the sensing tube of the thermostat in cooler air, satisfied and breaking the circuit to the compressor. Drawing B shows the thermostat in a closed position, calling for cooling and completing a circuit to the compressor. A thermostat of this type uses a chemical vapor in the sensing tube that reacts to temperature change. As the temperature drops, the pressure drops, and this allows the spring pressure to overcome the bellows pressure; this in turn separates the contact points inside the control. These contact points are connected to male electrical connections that exit the control body.

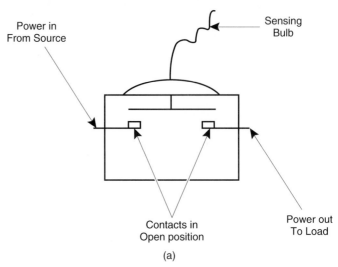

Figure 4-9 (A). When a refrigerator cold control is in an open position, no current will pass to the compressor.

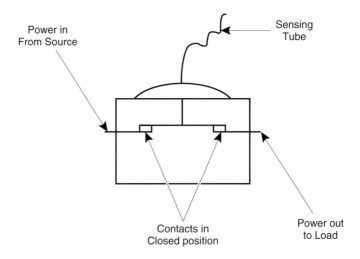

Figure 4-9 (B). When a refrigerator cold control is in a closed position, the compressor will be energized.

When the temperature rises, the pressure in the sensing tube increases. The bellows pressure then overcomes the spring pressure, causing the contact points to close again until cooler air causes the contacts to open again, until warmer air causes the contacts to close again, until ... thus the cycle of making and breaking a circuit to the compressor and accomplishing a temperature-controlled cooling cycle is established.

Adjustment by the customer to a different setting on the thermostat simply changes the spring tension, and a different temperature range is then achieved to cycle the unit off and on.

This same basic system, with some variations in design, is used on a significant number of oven thermostats installed by range manufacturers. An inexperienced person may perceive the sensing tube to be nothing more than a wire that may be cut or crimped, when in fact the opposite is true. The sensing tube is not a solid wire, but a hollow tube containing the temperature-sensitive vapor. Cutting or crimping the sensing tube damages the control.

4.6 BIMETAL THERMOSTATS

Although the bimetal thermostat also reacts to temperature, making and breaking an electrical circuit, it accomplishes the control function through a different method.

Figure 4-10 shows the method of operation of a bimetal thermostat. Two metal strips are positioned closely together inside the thermostat body. Each of the two metal strips inside the thermostat body is made up of two disimilar metals. The two metals are bonded together and, since they have different expansion ratios, the strip warps as the temperature changes. This warping, or bending if you prefer, causes the metal strips to make contact with each other. A thermostat of this type is popularly used in clothes dryers to sense temperature change, then energize or deenergize a heating element or gas valve. In this application, a control of this type is also used as a safety device, sensing excessive temperature in case of a blower motor failure, then breaking the circuit to the heat source.

Other uses for the bimetal type thermostat are as a defrost termination thermostat in a frost-free refrigerator, as a cycling thermostat in an ice maker or for a heating element in a dishwasher, and as an overload protector on a refrigerator compressor.

Bimetal thermostats are manufactured to fit many different applications. They may close on temperature rise, close on temperature drop, open on temperature rise, or open on temperature drop. Some function at a temperature of near 0° Fahrenheit, and some

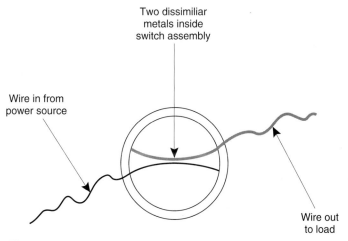

Two dissimiliar
metals inside
switch assembly

Wire in from
power source

Wire out
to load

Figure 4-10. When the two dissimilar metals inside a bimetal switch react to temperature, it makes or breaks a circuit to a load.

Defrost Termination
Thermostat

Refrigerator Thermostats

Laundry Equipment Cycling Thermostats

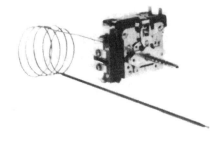

Oven Thermostats

Thermostats. Thermostats of different styles are used in appliances to control temperatures in refrigerators, laundry equipment, and cooking equipment. *(Courtesy Gem Products, Inc.)*

operate near 200° Fahrenheit. Page 48 show some of the different sensing tube and bimetal thermostats used in appliances.

4.7 DOOR SWITCHES

A *door switch* may be used to turn a light off and on in an oven or a refrigerator, to turn on a fan in a refrigerator when the door is closed, or in microwave ovens or laundry equipment to prevent operation of the unit when the door or lid is open.

Door switches are manufactured in three basic configurations: plunger style, rocker style, and interlock style. Page 50 shows the types of switches used in various applications on refrigerators, cooking equipment, and laundry equipment.

You will often encounter the terms *normally open* (*NO*) and *normally closed* (*NC*) when switches are discussed in manufacturer's service manuals or shown on wiring diagrams. These terms refer to the electrical position of the switch when its activating lever or mechanism is not being affected by a door or bracket. A switch that operates a fan in a refrigerator, for example, would be termed a normally open switch, because, when the door is opened, the switch would be at its "at rest" position and the contact points would be open, breaking the circuit to the fan.

A refrigerator light switch, though, would be exactly the opposite. Its "at rest" position, unaffected by the pressure of the door would be normally closed (NC) because, when it is not being contacted by the open door, the contacts inside the switch are closed and the bulb is energized through the complete electrical circuit. When the door is closed, then, the switch contacts are opened and the circuit to the light is broken. Page 50 shows some of the switches used in refrigerators, washing machines, dryers, dishwashers, and cooking equipment.

4.8 TIMERS

Timing devices are used in laundry equipment, refrigerators, dishwashers, and cooking equipment. They range in complexity from a simple mechanical timer that winds down to zero from a preset point to a solid state printed circuit board assembly that is touch-sensitive.

Inexperienced persons often incorrectly diagnose the timer as being the problem in an inoperative appliance because they view the component as being much more complex than it is. An electromechanical timer, such as one commonly used in an automatic washer, is only a series of switches that are either open or closed depending on the position of the cam at any given time during the cycle of the component. A *cam* is an assembly within a timer. It has raised areas that push on contact points and complete a circuit to

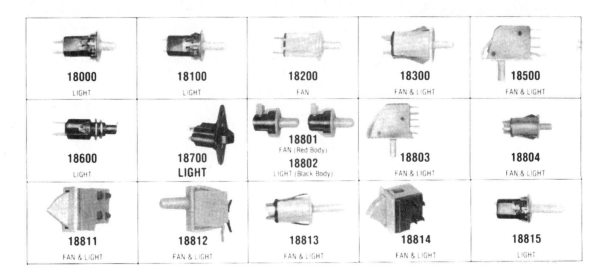

Refrigerator Door Switches

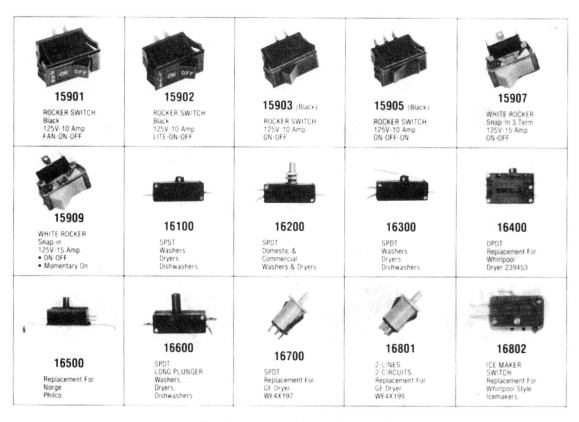

Switches In Laundry And Cooking Equipment

Appliance switches. Switches are used in appliances to control the operation of lights, fans, and other functions. *(Courtesy Gem Products, Inc.)*

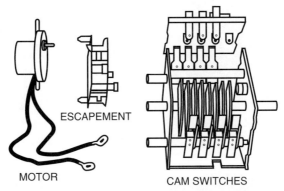

Figure 4-11. A washing machine timer. When the motor turns the escapement, the leaf switched inside the timer make and break a circuit to various components within the machine.

various other components within the appliance. Figure 4-11 shows the basic construction of an automatic washer timer. A single wire will bring the power into an automatic washer timer, but that power will leave the switching assembly through several different paths and energize several different components such as mixing valves, drive motors, fabric softener dispensers, transmission shifting solenoids, or brake assemblies.

When used as a control in a frost-free refrigerator, the timer has two cycles, run and defrost. Referred to as the defrost timer by many manufacturers, it maintains a run cycle most of the time, then enters the defrost cycle for a comparatively short period of time. Depending on the make and model of the refrigerator, defrost cycle lengths may vary from a minimum of 18 minutes to a maximum of 28 minutes. Page 52 shows several variations of refrigerator defrost timers.

A wide range of timers are used in clothes dryers, dishwashers, gas and electric ranges, and microwave ovens. Timers, their specific application, and method of operation will be covered in later chapters.

4.9 MOTOR STARTING RELAYS/OVERLOAD PROTECTION DEVICES

A *relay* is a device that is used in an application where a particular style of motor is used in an appliance. A refrigerator compressor motor, for example, is made up of a start winding and a run winding. It requires a relay assembly to energize both windings at the instant of start, then maintain power to the run winding for the remainder of the run cycle. You'll also encounter starting relays on some dishwashers and certain makes (notably General Electric and Hotpoint) of automatic washers.

Overload protection devices are designed to protect the motor in the event of an overheat or improper voltage situation. Motors used in appliances will be equipped with either internally or externally

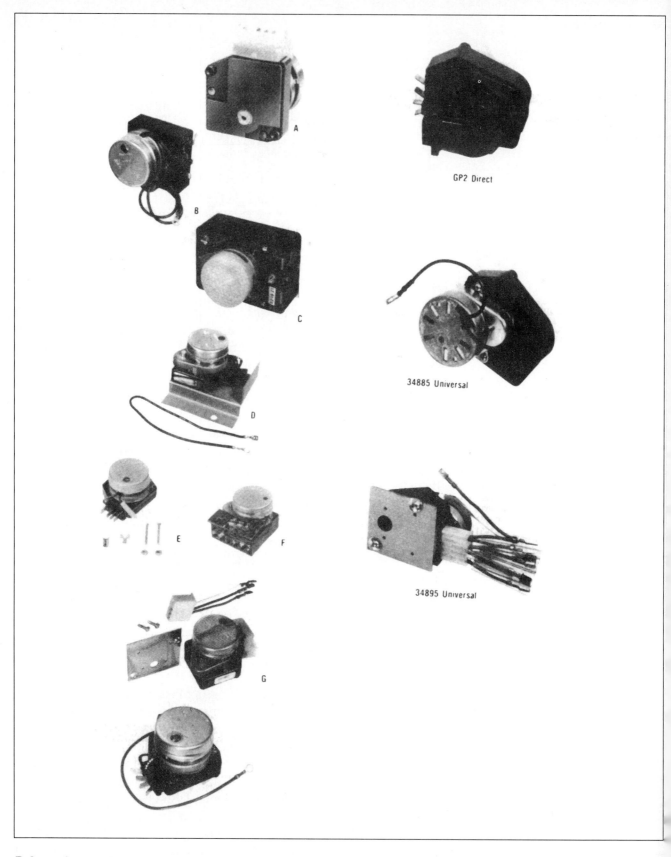

Defrost timers. While all defrost timers perform the task of initiating and terminating the defrost cycle, many different styles of timers are used in different brands of refrigerators. *(Courtesy Gem Products, Inc.)*

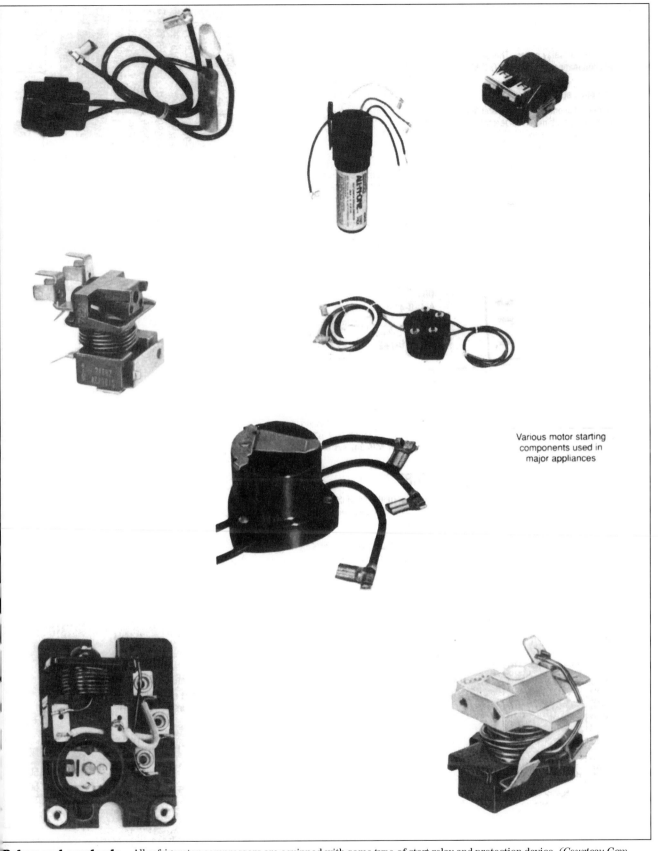

Various motor starting
components used in
major appliances

Relays and overloads. All refrigerator compressors are equipped with some type of start relay and protection device. *(Courtesy Gem Products, Inc.)*

mounted protection devices that react to excessive temperature or amperage draw. Page 53 shows some of the relays, overloads, and relay/overload, combinations used on refrigerator compressors. Overload protection devices and their uses in specific applications will be covered in detail in later chapters.

CHAPTER FOUR   **SUMMARY**

Electrical power is applied to a wide range of heating elements, motors, and lights in the operation of household appliances. Motors are used to operate automatic washers, automatic dishwashers, and clothes dryers, as well as cool compressors, and to circulate cool air inside a refrigerator cabinet.

An electric motor operates on magnetism and the law of electrical charges that states like charges repel and unlike charges attract. The electromagnetic field set up in the stator windings of the motor causes the magnet in the rotor section to spin when affected by the north and south poles of the magnets. Once a motor has begun to operate, the push and pull of the magnets and the centrifugal force results in a continuous spinning of the rotor.

Depending on the application of the motor, a start capacitor may be used to provide extra starting torque, such as in the case of an automatic washer, or a run capacitor may be wired into the motor circuit to allow a refrigerator compressor or window air conditioner fan motor to operate more efficiently. When a run capacitor is used, the motor is referred to as a *PSC (permanent split capacitor)* motor. Small shaded pole motors with very low starting torque are used to circulate air in refrigerators.

Electrical energy is used to operate water valves used in major appliances. A solenoid uses an electromagnetic field to cause a rod to change position within a valve assembly and overcome the spring pressure that holds the rod in place. This creates an opening that allows water to flow through the valve.

A single-solenoid valve is used to allow water flow into an ice maker in a refrigerator and a single-solenoid valve may also be used on an automatic dishwasher. Double-solenoid valves may be found in some makes of dishwashers. When used in an automatic washer, the double-solenoid valve is known as a *mixing valve* that allows the customer to select cold, warm, or hot water for the wash and rinse cycle. A double-solenoid valve used in a refrigerator equipped with an ice maker and a chilled water feature will have one inlet and two outlets for water supply.

Solenoids may also be used to operate locking devices on self-cleaning ranges or on door locks on some models of microwave ovens, or to operate fabric softener or bleach dispensers in automatic washers. A solenoid may also be used as an off-balance switch in an automatic washer or to operate a buzzer in a clothes dryer.

Heating elements are used in cooking and laundry equipment and to heat water in an automatic dishwasher. Bake and broil elements are made up of a resistive wire surrounded by a heat-conductive material that is wrapped in a metallic shell. This method of construction also applies to some defrost heaters used in frost-free refrigerators.

Another type of heating element is made up of a nichrome wire. Sometimes referred to as a *ribbon-type element*, this is used in clothes dryers and, when encased in a glass tube, is used as a defrost heater in a refrigerator.

Low-wattage heaters are also used in many different applications in major appliances. They may be a molded plastic heater, stranded heater wire wrapped in what appears to be standard wire insulation, a braided metal element, or glued to a foil backing. These types of heaters are used as drain trough heaters, dew point compensators to prevent refrigerator cabinets from sweating; they may also be used as a heater around an opening for dispensing ice cubes.

Switches take many forms in major appliances. A switch is used to turn a light off and on, start and stop a motor, or energize and deenergize a heating element. A switch may take the form of a push button, a timing device, or a temperature-controlled switch. A cold control in a refrigerator and a thermostat in an oven are two types of temperature-controlled switches. They use a sensing tube that reacts to temperature and causes contacts inside the switch assembly to make or break.

A bimetal thermostat uses the principle of two dissimilar metal strips positioned closely together inside the switching assembly. When a change in temperature occurs, the two metals react, warping to either make or break the circuit to the load they control.

Timing devices are used in appliances to control the length of a cycle in washing and drying clothes or washing dishes, or to control the sequence of run and defrost cycles in a frost-free refrigerators. Timing devices range in complexity from a simple mechanical timer that winds down from a customer-chosen setting to an off position to a touch-sensitive printed circuit board. The electromechanical timer is used extensively in laundry equipment and as a refrigerator defrost timer.

A *start relay* is a device that is used to assist in the starting of electric motors used in major appliances. In many cases, a relay is used in conjunction with an overload protection device that reacts to excessive temperature or amperage draw, and breaks the circuit to the motor. Whatever the application, the start relay assists in starting an electric motor by momentarily energizing the start windings.

Using Electrical Meters in Appliance Servicing

LEARNING OBJECTIVES

After studying this Chapter, you will be able to:
1. Identify the different types of meters used in testing major appliance components and in troubleshooting circuits.
2. Use various meters to measure voltage, amperage, and resistance.
3. Identify specialty meters used in major appliance servicing.

■　■　■

An appliance service technician can comfortably run eight calls a day. In the course of that day, the chances of finding the same problem twice are slim. On one call, you may have to determine whether the voltage applied to the unit is correct. On the next call, you will have to test a control on a dead unit to find out if it is allowing current to pass, and you'll have to use an ohmmeter. Next, you may have to test a motor on an automatic washer to make sure it is drawing the correct amount of amperage. And, to round out your day, you may have to test an oven for the correct cooking temperature or troubleshoot the high-voltage section on a microwave oven.

To accomplish these tasks, you'll have to be familiar with the various types of meters used by technicians, and you'll also need a basic understanding of their method of operation. There are more different brands, types, and combinations of meters on the market than there are different brands of major appliances, and selecting the meters you prefer to use is a matter of personal preference and individual experience.

5.1　MEASURING VOLTAGE

Technicians frequently have to use a voltmeter (or the voltage section

Figure 5-1.   A typical analog-type meter used
by appliance technicians. Often referred to as a
VOM (volt/ohmmeter), it is used for measuring
voltage and resistance. *(Courtesy of Simpson
Electric Co.)*

on a multimeter) to test for proper voltage in a wall receptacle. Standard wall receptacles, such as the type you would plug a 120-volt lamp into or a 240-volt receptacle used for electric range and electric dryers, can be checked with a voltmeter. A voltmeter is also used when troubleshooting an appliance for improper operation. On a dishwasher, for example, that is not filling properly, a voltmeter can be used to find out whether or not power is being delivered to the solenoid of the water inlet valve.

A meter like the one shown in Figure 5-1 can be used for either purpose. A meter of this type is known as an *analog multimeter*. It is *analog* because it has a printed scale and because it uses a pointer that moves in response to the electrical input being measured. It is called a *multimeter* because, in addition to measuring voltage, it is capable of measuring other types of electrical inputs (values) such as resistance.

Your inclination might be to be intimidated by a multimeter when seeing it for the first time. There *are* several different scales printed on the meter face, a selector switch that determines the function being used, and, in many cases, several different input jacks into which your test leads may be placed, depending on which electrical function you are using. Don't be intimidated. All reputable manufacturers provide clear descriptions of a meter's features and a concise manual that explains how it operates.

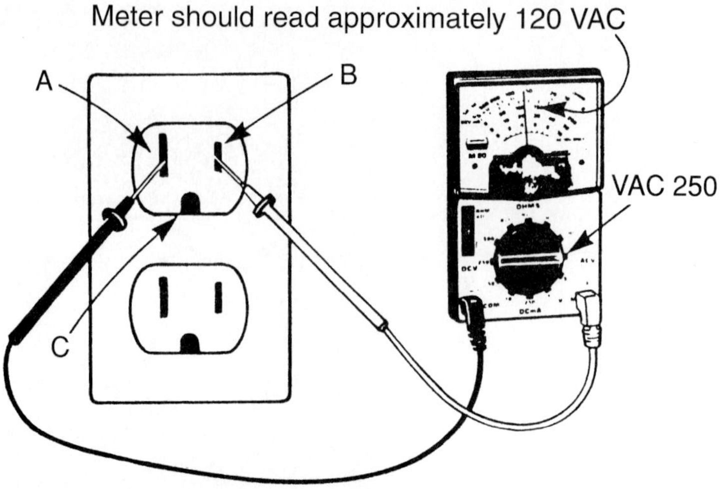

Meter should read approximately 120 VAC

A   B

VAC 250

C

Figure 5-2. Testing a 120-volt wall receptacle. On a good, properly wired receptacle, the readings will be 120 VAC between A and B, 120 VAC between B and C, and 0 volts between A and C. *(Courtesy Universal Enterprises, Inc.)*

The easiest test to learn is testing a standard wall receptacle. Following the manufacturer's instructions, place the test leads into the proper input jacks and set the meter to the proper function using the selector knob. A good rule of thumb to follow is to set the selector knob to the highest voltage scale available. Once you've established that you're within the operating range of the lower voltage scale, you can turn the selector knob to that scale. Figure 5-2 illustrates a wall receptacle test.

Caution! *When using a voltmeter to check for voltage, always check a known source first! This eliminates the possibility of being injured due to a false reading from a defective or improperly set-up meter!*

When using your multimeter to make this test, the manufacturer of your equipment may identify the proper setting with a *V* for volts or with a *VAC*, which stands for *volts alternating current*.

If using a meter is new to you, one simple way to overcome possible confusion in regard to the scales printed on the meter face is to measure a known quantity, such as a standard wall outlet, on several different scales. Then take your time and closely observe the position of the pointer when using the different scales available on your meter. This procedure will help you to become more familiar with your brand of equipment.

Another common wall receptacle test that appliance service technicians must perform is that of a 240-volt receptacle, into which an electric range or standard-sized electric clothes dryer is connected. This test is illustrated in Figure 5-3.

As a service technician, you'll also be required to determine whether a receptacle is wired properly in regard to polarity and safety grounding. When referring to "polarity," we mean the hot wire being connected to the hot side of the receptacle and the neutral wire being connected to the neutral side of the receptacle.

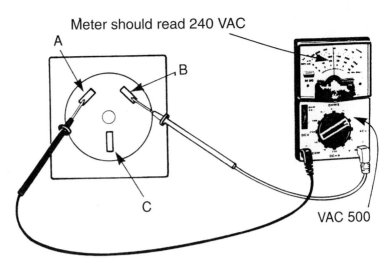

Meter should read 240 VAC

VAC 500

Figure 5-3. Checking an electric range receptacle. Your voltage readings should be as follows: 240 VAC between A and B, 120 VAC between A and C, and 120 VAC between B and C. *(Courtesy Universal Enterprises, Inc.)*

Why should this be of importance to you, the person who's there to troubleshoot and repair an appliance? Because an improperly polarized or improperly grounded wall receptacle can affect the operation of an appliance, particularly if that appliance contains any solid state controls or printed circuit boards.

Some technicians use a simplified tester to determine if an outlet is properly grounded and polarized. This particular tester is available at any good hardware store and uses a series of green and red lights to indicate if a receptacle is wired properly. Another method is to use your voltmeter.

This is illustrated in Figure 5-4. Our example shows a standard 120-volt wall receptacle. Upon close inspection of a standard wall outlet, you'll notice that one of the vertical slots is smaller than the other. In the case of our drawing, the smaller vertical slot is on the right and the larger vertical slot is on the left.

With a properly wired receptacle, the smaller slot will be wired to the hot (black) wire and the neutral (white) wire will be connected to the larger slot. This means that we will be able to use a voltmeter to read power at the proper place, and we will not read power at the improper places.

If you were to insert the probes from your voltmeter into the points marked A in the illustration, for example, you should read 120 VAC. You should also read 120 VAC at points C. A voltage reading at these two points on a receptacle indicates that it is properly polarized and wired correctly for safety ground. These readings are correct because you are reading the potential difference between the hot and neutral wires at points A, and you are reading between the hot leg and the ground at points C.

If you were to get a voltage reading at points A but not at points C, it would prove that your receptacle was not properly grounded.

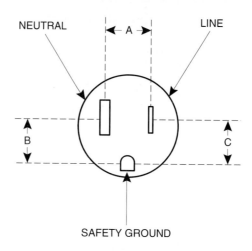

NEUTRAL   ← A →   LINE

SAFETY GROUND

A = 120 VAC B = 0 VAC C = 120 VAC	PROPERLY POLARIZED AND PROPERLY GROUNDED
A = 120 VAC B = 0 VAC C = 0 VAC	NOT GROUNDED
A = 120 VAC B = 120 VAC C = 0 VAC	PROPERLY GROUNDED BUT THE POLARITY IS REVERSED

Figure 5-4.   A receptacle that is properly wired for
correct polarity and proper ground will show a voltage reading at points A and C, but not at point B.
(Courtesy of White Consolidated Industries, Inc.)

Readings at points A and at points B would indicate that your receptacle was wired for safety ground but was not wired for proper polarity.

In accordance with manufacturer's specifications and for consumer safety, the only acceptable readings are at points A and
points C.

Another application in which you should be able to use your
voltmeter is in testing a *main control panel* or, as it is sometimes
referred to, *load center*. The load center in a residence or a commercial building contains the circuit breakers that provide branch circuits from the main power supply. In older homes you may find a
cartridge and plug fuse system, such as the one described in Chapter
Three. In any case, you should have a firm understanding of the load
center and how to test it for proper voltage.

As an appliance technician, you won't be called upon to replace
a defective circuit breaker. But to convince your customer that the
problem exists within the power supply system and not within the
appliance itself, you'll have to be able to prove it to their satisfaction.

Figure 5-5 shows a typical load center with circuit breakers and
a simple method of testing a single-pole breaker that provides a 120-
volt circuit.

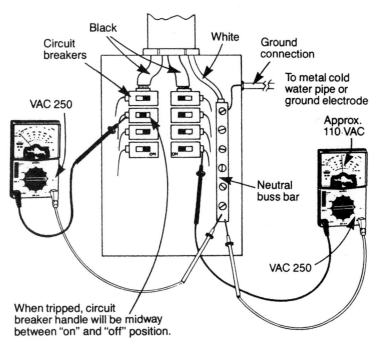

Figure 5-5. Testing a circuit breaker in a main disconnect panel. Touching the leads of the meter from either hot leg to neutral will offer a reading of 110 VAC. *(Courtesy of Universal Enterprises, Inc.)*

Caution! *Use extreme care! Whenever you remove a main panel cover to expose the main circuits for testing, you are at risk of electrical shock if you touch any of the wiring connections. Electrical shock can cause serious injury or death!*

In a 240-VAC, single-phase circuit, such as the one shown in Figure 5-5, you can use your voltmeter to test a circuit breaker to make sure it is providing voltage to a circuit. With your meter set to a range of 250 VAC or higher, touch one of the meter leads to the neutral bar inside the panel and the other meter lead to the wire connection at the circuit breaker. Since you are reading the potential from one hot leg to ground, you would read 120 VAC.

If you were to test a two-pole breaker, such as the one that would supply 240 VAC to an electric clothes dryer or electric water heater, you would read 240 VAC by touching your meter leads to the two wires connected to the breaker. As with the single-pole breaker, touching one lead of the meter to one of the hot wires and touching the other lead of the meter to the neutral bar would yield a voltage reading of 120 VAC.

5.2 MEASURING RESISTANCE AND TESTING FOR CONTINUITY

As mentioned previously, a meter popularly used by an appliance service technician is commonly referred to as a VOM or volt/ohmmeter.

In other words, in addition to testing for voltage, the meter can also be used to test a component, such as a bake element in an electric oven, for proper resistance. (You'll recall from Chapter Two that we identified the unit of measurement for resistance as the ohm.)

Depending on the type of component you're testing, it may have a resistance of 1 ohm (or possibly a fraction of an ohm such as in the case of run winding in a compressor motor). Or it may have a resistance of several thousand ohms, such as in a resistor in a printed circuit board. This wide range of variance is responsible for the different settings that exist on the resistance scale in a typical meter.

The meter you choose may have a selector that allows you to dial to several different resistance scale settings. They may range from an R×1 (read it *resistance times 1*) scale to an R×1,000 (read it *resistance times 1,000*) scale. The R×1 scale is commonly used by technicians to test the resistance of motor windings or coils used in solenoid valves, while the higher resistance scale is used to test capacitors or solid state components used in appliances. The lower ohms range is also used to test fuses and switches for continuity, to make sure that they will allow current to flow. The one thing you want to remember about using your meter to measure resistance or continuity is *never* use it on an energized circuit. *Always* unplug the appliance before using your meter in this mode. Failure to do so may cause damage to the meter.

The battery in your VOM is what does the work when the meter is used in this mode. The voltage from the battery supplies the energy that makes the needle rise. When you set your meter to the ohms range you've decided to use, test the ability of the meter by touching the two meter leads together. If your battery is in good shape, you should be able to "zero" the meter. This means you are able to use the ohms adjustment knob to park the indicator needle directly on the zero, which in a typical meter is located at the extreme right of scale.

Figure 5-6 shows this procedure. After plugging the test leads into the meter at the proper locations, you will be able to cause the pointer to rise and swing to the right by touching the free ends of the

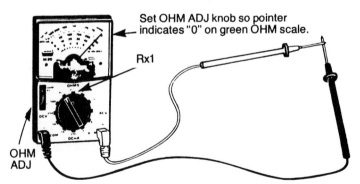

Figure 5-6.   Preparing to use an ohmmeter. By touching the test leads together, the meter can be adjusted to read 0 resistance. (*Courtesy of Universal Enterprises, Inc.*)

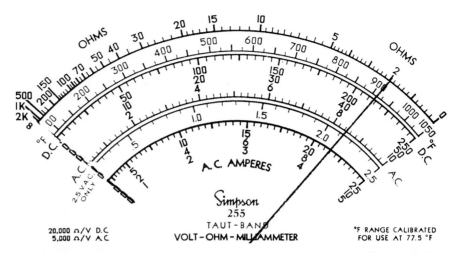

Figure 5-7. When reading 2 ohms resistance to current flow, the pointer will not go all the way to a reading of 0, but will instead swing to the right only as far as the number 2. *(Courtesy of Simpson Electric Co.)*

test leads together. To zero the meter on the R×1 (resistance times 1) scale, select the R×1 scale, touch the two leads of meter together, and turn the ohms adjustment knob until the needle settles on the zero at the far right of the scale.

Once you have accomplished the fine tuning of your meter on the ohms Rx1 scale, you can test a component for proper resistance. Let's say, for example, that you wanted to confirm that a run winding in a compressor does, in fact, have a resistance of 2 ohms. (Refer to Figure 5-7.) As you can see from the typical meter face, such as the one shown in Figure 5-7, the ohms scale is the uppermost scale. Touching the two test leads of the meter together would allow your indicator to point to the 0 at the far right of the scale. Touching the two test leads of the meter to the electrical connections of the motor winding, then, would offer a reading of 2 ohms. In other words, when you touch the two leads of the meter together, you are reading a resistance of 0, because the pointer swings all the way to the right. But when you touch the meter leads to the electrical connections of the motor, the pointer will not go as far because it is now reading the resistance of the motor winding: in this case, 2 ohms.

Switching to the R×100 scale would enable you to read a higher resistance. This is shown in Figure 5-8. In this case, the pointer is still pointing to the number 2 on the scale but with the meter set at R×100, the 2 now represents a reading of 200 ohms.

Note that you can also read 200 ohms on the R×1 scale in the meter face shown in Figure 5-8. On the R×1 scale, the pointer would rise only a very short distance to the number 200 located near the far left of the scale.

Testing for continuity is also accomplished with the meter set on the R×1 scale. Simply defined, the term *continuity* means that there is no break to prevent current from flowing through a wire, a

Figure 5-8.   A meter reading 200 ohms resistance. The pointer still indicates the number 2, but the scale has been changed by turning the function knob of the meter to the R×100 scale. *(Courtesy of Simpson Electric Co.)*

fuse, or the contact points on a switch. Continuity testing also applies to testing a wiring harness. If a wire has continuity, it will allow electrical energy (current) to flow. If a wire doesn't have continuity, a complete circuit won't exist. Figure 5-9 shows some of the continuity tests that are accomplished with an ohmmeter.

   Remember: *When checking continuity on a component, disconnect the component from the circuit. Failing to do so could give you a false continuity ("good") reading.*

   A term you will become familiar with when using an ohmmeter according to manufacturer's instructions is *infinity*. This term is identified by manufacturers with a symbol that appears to be a figure eight lying on its side (∞). By description, it applies to the term *infinite resistance*, meaning that there is a total resistance to current flow in the electric circuit. A switch that is in the off position or a fuse that is blown will offer a reading of infinity (or open) when the leads of the meter are applied, as shown in the illustration.

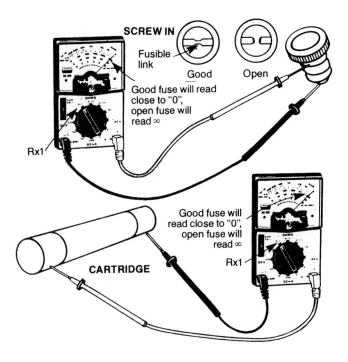

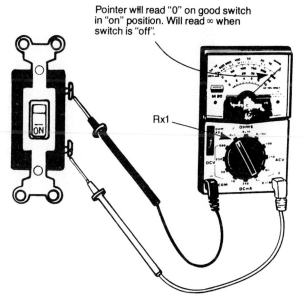

Figure 5-9. Using an ohmmeter on the R×1 scale to test for continuity of fuses and a switch. Some meters are equipped with a continuity setting on their function knob, and may also use an audible tone to indicate continuity. *(Courtesy of Universal Enterprises, Inc.)*

5.3 MEASURING CURRENT DRAW IN A CIRCUIT

An appliance service technician will often use an ammeter while troubleshooting a problem with an appliance. The function of an ammeter is to measure the current (amperage) draw in a circuit.

Once a technician has established the proper current draw of a component, such as an automatic washer motor, the ammeter can be used to test the component during operation to determine whether it is defective.

Ammeters, such as the one shown in Figure 5-10, are often referred to as *clamp-on testers*, because of the design of the meter and the method by which it is used. The jaws of the meter are spread apart by overcoming a spring pressure that holds them together and, once they are opened, they are clamped around a conductor. Through this method, the ammeter can measure the current flowing in a circuit.

The important thing to remember about using an ammeter is that it must be clamped around a single conductor to do its job. You could not, for example, clamp it around a 120-volt power cord connected to a refrigerator and obtain a reading. You would have to split the cord so that the ammeter was clamped around only one conductor, not the hot and neutral wire together.

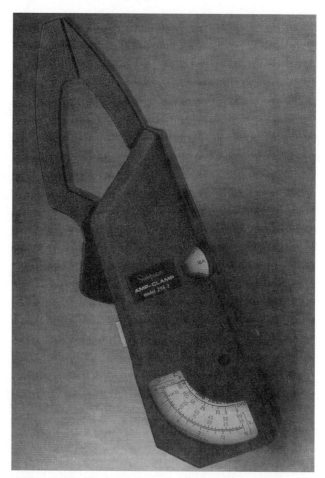

Figure 5-10. An example of an ammeter used by an appliance technician. The jaws of the meter are clamped around a wire to allow the meter to read the level of current draw flowing in a conductor.

An ammeter may also be used to test the operation of 230-volt appliances, such as clothes dryers or electric ranges, but, as in a 120-volt circuit, the meter must be clamped around only one conductor. Clamping it around the two hot wires that are sealed together within the molded power cord for such an appliance would not allow the meter to read the current flowing in the circuit during its operation.

5.4 THE VOLT/WATTMETER

Some appliance technicians use a volt/wattmeter to test an appliance for proper current draw. The volt/wattmeter differs from the clamp-on style ammeter in that the appliance is unplugged from the wall outlet, then plugged into the volt/wattmeter. The meter itself is then plugged into the wall outlet. When the appliance is operated in this manner, this combination meter indicates the voltage being applied on one meter face, while a second meter face indicates the wattage draw of the operating appliance. The volt/wattmeter is equipped with a standard power cord, such as that found on a refrigerator, and is usually used to test only those appliances that operate on 120 VAC.

You'll notice that we've used a different term when discussing the volt/wattmeter: *wattage*. Amp draw and wattage are fundamentally the same concept, and so there is a relationship between the terms *watts* and *amps*. You can refer to Chapter Two for a review of current draw in a circuit: How the wattage draw of a component is calculated by first measuring amperage and then multiplying it by the voltage applied to the appliance to calculate the wattage.

As we said at the beginning of this chapter, an appliance technician may use several different meters during the course of day. Thus far, we have discussed only a few of those test instruments that accomplish the testing of capacitors, testing for leakage around a microwave oven door, or testing for proper operating temperature of a conventional cooking appliance. More of these specialty meters, as well as other specialty tools, will be discussed as we cover specific service procedures in later units.

CHAPTER FIVE   SUMMARY

During the course of a day, an appliance technician may use several different meters to solve various problems in different appliances. A popular meter used is the *analog multimeter*, so-called because it uses a pointer that moves in response to an electrical input. A multimeter may be used to measure the available voltage in a wall outlet, to measure resistance in a circuit, or to test the continuity of a switch.

The multimeter may also be used to determine whether a receptacle is wired for proper polarity. This is important to understand because an improperly wired circuit can cause problems with the operation of an appliance, particularly if the unit uses printed circuit boards in its operation.

On occasion, a technician may be required to test a circuit breaker in a main disconnect panel to prove to a customer's satisfaction that the problem with their appliance is in the wiring circuit and not in the appliance itself.

Caution! *Always use extreme caution when conducting electrical tests within the main circuit panel. Removing the cover exposes hot wires, and touching any of the wiring connections could result in serious injury or death from electrical shock.*

Current draw in a circuit is measured with a clamp-on meter known as an *ammeter*. The ammeter is designed to read current flowing in a circuit (the amperage draw of the load) and can only work properly when it is clamped around a single conductor.

Another meter used by appliance technicians is the *volt/wattmeter*. This is a combination meter that is plugged into the 115-VAC wall outlet with the appliance plugged into the meter. One meter face indicates the voltage being applied to the appliance, and the other meter face indicates the wattage draw of the appliance.

Appliance technicians use a wide variety of meters to test for voltage, amperage, microwave oven leakage, temperature testing, and component testing. Many different types of meters, such as analog- or digital-type, are available to the technician. The selection of meters is a matter of personal preference and individual experience.

Refrigeration Fundamentals

LEARNING OBJECTIVES

After studying this Chapter, you will be able to:

1. Describe the method by which the refrigeration process takes place in a refrigerator.
2. Identify the components in a refrigeration system.
3. Trace the direction of refrigerant flow through a system.

■ ■ ■

There are those who argue that an appliance technician's job is to repair a refrigerator, not design one. That being the case, it's not necessary to study the methods of heat transfer (often referred to as the *laws of thermodynamics*) to be able to repair a refrigerator that's not functioning properly. While we're in agreement that extensive study in math and engineering is not necessary to become a competent technician, a firm understanding of the basics of how we move heat out of a refrigerator cabinet is essential. Without this understanding, technicians can function to a certain degree. With it they can function with far more confidence and, as a result, be not just competent, but excellent.

Heat is a form of energy that cannot be destroyed, but it can be transferred from place to place. The fundamental definition of *refrigeration* is the transfer of heat from a place where it's not wanted to a place where it's not objectionable. From your point of view as an appliance technician, the place where the heat is not wanted is inside the refrigerator; the place where it's not objectionable is outside the cabinet. As we go through this unit, keep one fundamental idea in mind: *We don't put the cold into a refrigerator, we take the heat out.*

6.1   HEAT TRANSFER

Some basic laws of nature govern the transfer of heat. The first one to be aware of is that heat always moves from a warmer surface to a cooler surface. You could use some complex physics terms to describe this process if your aim was to impress someone, but since our goal is simply to make you understand how a refrigerator works, we'll take the simple approach. If you were to park a black automobile in the Arizona desert in the month of August, the temperature of the hood of that car would probably rise to about 125° Fahrenheit. If you were to then place your hand on the hood of that car, you would definitely experience the transfer of heat from a warmer surface to a cooler surface. Why? Your body temperature is normally 98.6°, and, since you are cooler than the hood of the car, the heat would begin to transfer from the warmer hood to your cooler hand.

To further illustrate this concept of heat transfer, imagine that the same automobile is parked in Minnesota in the month of January. If you were to place your hand on the hood of the car in this location, you would again experience the transfer of heat from a warmer surface to a cooler surface, except this time the heat would be moving from your body to the car.

Another law of nature that governs the transfer of heat is that heat may move in three ways: radiation, convection, and conduction. Understanding each of these will make it easier for you to eliminate the "mystery" behind the process of refrigeration.

Radiation   Formally defined, *radiation* is the movement of heat through the air, which doesn't heat the air, but instead heats solid objects. The heating of these solid objects will, in turn, heat the surrounding area. The simplest illustration of radiation is heat energy from the sun. This heat energy passes through the atmosphere without warming it up. But, as the heat energy comes into contact with the earth, it warms its surface. This is, then, one method by which the heat we need to take out of the refrigerator gets there in the first place. It *radiates* in when the cabinet door is opened.

Convection   The movement of heat through a fluid is called *convection*. This term is important in the explanation of refrigeration fundamentals for two reasons: First, it explains another method of heat movement into the refrigerator cabinet (moisture in the air carries some of the heat). Second, it explains why a liquid, such as a pitcher of juice, is chilled when it is placed warm into the refrigerator and is cooled by the surrounding air. Convection is also at work when heat is transfered into or out of the fluid that we use inside the coils of the refrigeration system in the refrigeration process.

Conduction   The movement of heat through a solid material is known as *conduction*. This is best illustrated by putting one end of a metal rod into a campfire. If you were to leave the metal rod in this position for several minutes, then touch the end of the rod that was not in the fire, you would feel the heat from the fire because it traveled through the solid material from the contact end to the end you touched. Conduction is the process by which the heat moves through glass or metal containers into the cooler surrounding air in the refrigerator, and also explains how the heat moves through the metal tubing that makes up the refrigeration system coils.

6.2 HEAT ABSORPTION AND REJECTION

There are two other basic laws of nature to consider when explaining how a refrigeration system works. They are:

1. When a substance boils, it absorbs heat.
2. When a substance condenses, it rejects or, if you prefer, releases heat.

These two laws relate to what we call a *change of state*, which is the condition that occurs when a substance changes from one physical state to another, such as from a liquid to a vapor. Specifically, the idea of heat absorption and rejection (or dissipation) relates to the change of state that occurs in the fluid contained inside the refrigeration system tubing. Changing in state from a liquid to a vapor allows for the absorption of heat, while a change in state from a vapor to a liquid allows for the rejection (dissipation) of heat.

6.3 REFRIGERATION SYSTEM COMPONENTS

The system that performs the job of preserving the food in a refrigerator is made up of four components: (1) compressor, (2) condenser, (3) evaporator, and (4) metering device. There are some fundamental differences to be found in the method of construction of some of these components depending on the specific type of refrigerator you are servicing. We'll discuss these differences in later units. At this point, the basic definitions, of these components will suffice.

Compressor A pump that is designed to accept a vapor at a lower pressure, and then discharge that vapor at a higher pressure, is called a *compressor*. The compressor is necessary because, in addition to the required change of state of the fluid inside the system coils, a pressure differential is also necessary for a refrigeration system to function. The compressor is the component that allows one section of the refrigeration system to be the high-pressure side and the other section of the refrigeration system to be the low side.

Condenser   The *condenser* is the section of the refrigeration system located on the outside of the cabinet, allowing for the rejection of heat into the room. It is located on the high-pressure side of the system, which would mean that the compressor discharges into condenser.

Evaporator   This component is the section of the refrigeration system located on the inside of the refrigerator cabinet. It allows for the absorption of heat and is located on the low-pressure side of the system. From the point of view of sequence, the evaporator is "ahead" of the compressor. This means that the compressor draws low pressure from the evaporator, then discharges into the condenser.

Metering Device   A metering device is the section of the refrigeration system that separates the high-pressure side from the low-pressure side. Specifically, the metering device in a refrigerator is referred to as a *capillary tube*, which is tube of a given length that has a very small diameter opening. This small diameter opening, combined with the length of the tube, allows for a pressure drop from the high-pressure side of the system to the low side. The capillary tube is located between the condenser and the evaporator.

6.4   REFRIGERANT

Considered by some to be the "fifth component" in the basic refrigeration system, the refrigerant is the chemical compound that actually does the work and allows for the transfer of heat from the inside of the refrigerator cabinet to the outside of the cabinet. Known commonly as *Freon*, which is actually a brand name, the refrigerant popularly used for many years in refrigerators is known as R-12. *R-12* is a chlorine-based chemical that has a very low boiling point, –21° Fahrenheit. Since R-12 is a chlorine-based chemical, it has been targeted by the Environmental Protection Agency as a substance that is causing damage to the ozone layer above the earth, and, as such, must be phased out of production. An alternate refrigerant designed to be safer for our environment and known as *HFC-134a*, has been phased into use for domestic refrigerators. We'll discuss alternate refrigerants and laws covering the use of refrigerant recovery systems in Appendix A.

6.5   THE REFRIGERANT CYCLE

With the basic laws of heat transfer and the identification of the components of a refrigeration system accomplished, it's time to put all this information together. In simple terms, the "refrigeration" sequence is to absorb the heat from the food or drink inside the cabinet and then dispense it to the outside of the cabinet. Using a drawing that shows the system components, we will identify the change in state of the refrigerant and the change in pressure in the refrigerant cycle. This is your first opportunity to become involved in an interactive learning process with this textbook. To get the maximum benefit from this exercise, you'll need four colored pencils:

Red: Color code for high-pressure vapor.

Purple: Color code for high-pressure liquid.

Green: Color code for low-pressure liquid.

Blue: Color code for low-pressure vapor.

As we progress through our explanation of the sequence of the refrigeration cycle, use the colored pencils to code the state of the refrigerant as it travels through the system components.

You may use any pattern you like to indicate the refrigerant as it is contained in the tubing assemblies. But keep in mind that you will be mixing two colors in several instances to indicate the change in state taking place inside a component. In the condenser, for example, the refrigerant enters as a high-pressure vapor (color code red), and, as it travels through the condenser, it changes in state from a vapor to a liquid. This means that through the middle of the condenser you'll have to show some high-pressure vapor and some high-pressure liquid to be correct in your illustration of what's going on inside the condenser.

Illustrating the Refrigeration Cycle and Condition of the Refrigerant as it Travels Through the System

Refer to Figure 6-1. Locate the compressor near the bottom of the refrigerator cabinet. Use a red pencil to indicate a hot, high-pressure vapor leaving the compressor through the discharge line and entering the condenser.

The color code at the inlet of the condenser may be shown totally red to indicate that the condition of the refrigerant at this point is all vapor. As you code the condenser, however, begin to use the purple to show a beginning of the change in state from a vapor to a liquid. At the midpoint of the condenser, you should be showing an even mixture of red and purple in color coding. Near the end of the condenser, the majority of the color code will be purple, with a small amount of red. The liquid line, which is the tube that connects the end of the condenser to the capillary tube, will be all purple, the color code showing high-pressure liquid.

Continue with the purple color through the capillary tube, changing to the color code green near the end of the capillary tube to indicate a low-pressure liquid. At the inlet of the evaporator, the color

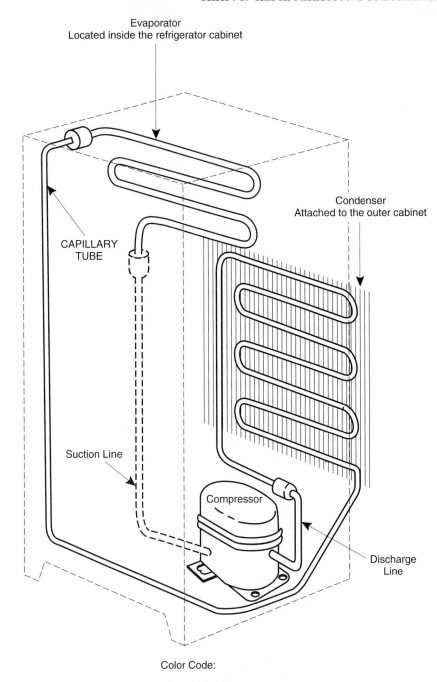

Color Code:

Red: High pressure vapor
Purple: High pressure liquid
Green: Low pressure liquid
Blue: Low pressure vapor

Figure 6-1. An illustration of the basic components in a conventional refrigerator/
freezer. You may use colored pencils to code the condition of the refrigerant as it is
pumped through the sealed system.

code will be almost all green, with small amounts of blue to indicate
the onset of the change in state to a vapor.

Midway through the evaporator, your color code should show a mixture of green and blue. Near the end of the evaporator, the color code will be almost all blue to show mostly vapor with a small amount of liquid.

The color code in the suction line will be all blue to indicate a low-pressure vapor state as the refrigerant returns to the compressor. Your color coding of the illustration should now be complete.

Now let's review the refrigeration cycle and change in state of the refrigerant in more detail. Refer to your color-coded illustration as we outline the method of operation. Our starting point is at the inlet of the evaporator, as the refrigerant enters as a low-pressure liquid (code green).

Recall one of the basic laws of heat transfer governing the operation of a refrigeration system: As a substance boils, it absorbs heat. This law comes into play in this section because the refrigerant is evaporating (changing from a liquid to a vapor) and, in the process of doing so, absorbs heat. Also a factor is the heat transfer law that says, heat always moves from a warmer surface to a cooler surface. This is because the heat from the air, which entered either through the opened cabinet door or with the warm items placed into the refrigerator, will move toward the cooler surface of the evaporator.

This heat will move through radiation and convection until it contacts the metal tubing, at which time it will move by conduction through the solid material, ultimately being absorbed into the evaporating refrigerant inside the tubing assembly.

One thing that should be clarified is the use of the term *boiling*, when referring to the function of the refrigerant as it changes in state from a liquid to a vapor, absorbing heat in the process. The average person is conditioned to respond to the word *boiling* by recalling that water boils at 212° Fahrenheit. From this point of view, confusion could result when trying to understand how something could be boiling, yet at the same time be cold and covered with frost. (In fact, most refrigerator evaporators can be found to be at a temperature near –10° Fahrenheit when the system is operating.) How can this be? The chemical compound used in a refrigeration system simply has a boiling point that is much lower than that of water, and as a result, the boiling (evaporation) takes place at this lower temperature.

Once the refrigerant has changed totally in state from a liquid to a vapor (code blue) and done its job of absorbing heat, the refrigerant now travels down the suction line, into the compressor. The compressor is one of the components providing the pressure differential necessary for the operation of a refrigeration system.

The refrigerant is heat-laden as it exits the compressor as a high-pressure vapor (code red). As it enters the condenser, another heat transfer law that applies: When a substance condenses (changes in state from a vapor to a liquid), it dissipates, or gives up, heat. In this case, the heat given up by the refrigerant is the heat that was absorbed in the evaporator. The heat will leave the condenser coil through conduction and pass into the air surrounding the condenser. The reason is that the condenser is warmer than the ambient air temperature and "heat always moves from a warmer surface to a cooler surface."

The refrigerant now travels through the liquid line (code purple) and moves to the capillary tube, the system's metering device. This small diameter tube is the other component that provides for a pressure differential (in this case a pressure drop) necessary for the operation of a refrigeration system.

The capillary tube is sometimes referred to as an *expansion device*, because the pressure drop created by the metering device is what fundamentally allows for the expansion of the refrigerant as it enters the evaporator and begins to change in state.

By the time the refrigerant has traveled through the capillary tube, it is now a low-pressure liquid (code green). As it enters the evaporator, our cycle repeats.

CHAPTER 6   SUMMARY

To be an effective appliance service technician, you need a firm understanding of the fundamental laws of heat transfer governing the operation of a refrigeration system. The laws of heat transfer are:

1. Heat always moves from a warmer surface to a cooler surface.
2. Heat moves in three ways: radiation, convection, and conduction.
3. When a substance boils, it absorbs heat.
4. When a substance condenses, it rejects (gives up) heat.

The four basic components of a refrigeration system are the compressor, condenser, evaporator, and metering device. The evaporator is located on the low-pressure side of the system, and heat is absorbed by the refrigerant as it changes from a liquid to a vapor. The condenser is located on the high-pressure side of the system, and heat is rejected by the refrigerant as it changes in state from a vapor to a liquid. The compressor and the metering device provide the pressure differential necessary for the operation of a refrigeration system.

The definition of the *refrigeration process* is the transfer of heat from a place where it is not wanted to a place where it is not objectionable. In the operation of a refrigerator, the system does not put the cold in; it takes the heat out.

Refrigerator/Freezer Servicing

LEARNING OBJECTIVES **After studying this Chapter, you will be able to:**
1. Identify the different types of compressors, condensers, and evaporators used in various types of refrigerator/freezers.
2. Use a wiring diagram to identify components and trace the electrical circuit of a refrigerator/freezer.
3. Describe the different methods that manufacturers use in the construction of various styles of refrigerator/freezers to accomplish air flow and food preservation.

■ ■ ■

If an appliance service technician makes ten calls in one day and works on nothing but refrigerator/freezers, the chance that they would all be the same in construction, specific method of operation, and style would be, to say the least, slim. As a technician, you have to contend with the fact that several different manufacturers of appliances are in competition with each other. As a result, not only do differences exist from one company to another, but manufacturers also have to change their style of construction from one year to the next to maintain their position in the industry. What this means to you is that the 1982 Admiral side-by-side you repaired at 9 o'clock in the morning doesn't look the same as the one you'll be troubleshooting at 4 in the afternoon.

There are some basics that manufacturers can't change. But, to keep the new models coming year after year, a panel may be removed in a different way, a fan motor might be mounted differently, or a feature that appeared last year won't appear on this year's model. Put this all together with the fact that all manufacturers have a product line that ranges from a "bottom-of-the-line" apartment-style refrigerator/freezer to a 28-cubic-foot side-by-side with an automatic ice maker and a chilled water dispenser, and you've presented a challenge to the person who services appliances. Not only must you become familiar with what is already out there, but you must also keep up with the new developments that are bound to occur.

Studying a textbook or attending a trade school or college training program will give you the background you need to begin working with refrigerator/freezers. The thing to realize, however is that, for you to deal with the various situations as they arise, there's no substitute for field experience.

7.1   TYPES OF REFRIGERATOR/FREEZERS

Whether it's built by Whirlpool, Amana, WCI, or General Electric, there are three basic refrigerator/freezer categories of construction: conventional, cycle defrost (sometimes referred to as automatic and therefore confused with frost-free units), and frost-free. In addition to the three basic categories, there are also variations in cabinet styles, such as single-door, top-mount (a manufacturer's way of saying that the freezer is on the top and fresh food compartment on the bottom), bottom freezer, and side-by-side. For these different types of units to operate, variations of sealed system components and electrical systems are used.

7.2   CONVENTIONAL REFRIGERATOR/FREEZER

The *conventional unit*, so named because it is the simplest of all refrigerator/freezers, contains the sealed system components—the compressor, condenser, and evaporator—in their simplest form without any forced air or defrost systems. Its cabinet is known as *single-door construction* because there is no separate freezer section door outside the cabinet, only an inner freezer door. This type of unit is often found in apartments and may also be supplied with mobile homes. They can be as small as 9 cubic feet in storage capacity and may range in size up to 12 cubic feet. As a general rule, anything from 13 cubic feet on up will be a two-door model unit.

The Basic Refrigeration Unit

An illustration of the conventional refrigerator sealed system is shown in Figure 7-1. The picture shown is actually a page out of an Admiral service manual from a 1976 model unit. Before you jump to the conclusion that we're avoiding using state-of-the-art information, consider what we said at the beginning of this chapter: you need to be familiar not only with the new items that are coming out, but also with what is already out there. In addition, we chose this illustration

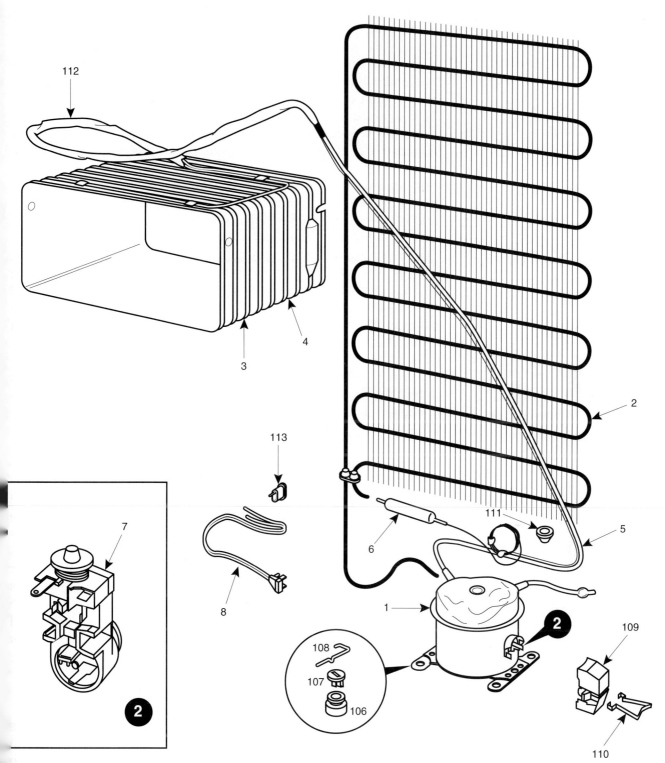

Figure 7-1. Sealed system parts of a conventional refrigerator shown in a reproduction from a manufacturer's service manual.

because it shows the basic refrigeration system as it was used then and is being used in manufacturing today. The basic refrigeration system used in this type of unit is just that—the basic refrigeration system—and there's nothing that can be done to change the fundamental method of operation. The basic idea of using the evaporator section of the system to form the freezer section and allowing the cold air to fall by *natural convection* down to the fresh food section below was effective in 1976 and is still effective today. The manufacturer's goal in any refrigerator/freezer, whether it's 20 years old or brand new, is to maintain the desired temperature in the freezer section (approximately 10° Fahrenheit in the case of this type of unit), and in the fresh food compartment (approximately 40° Fahrenheit). In the case of the conventional refrigerator, the freezer section is really nothing more than just a cold chest within the main cabinet. Two-door refrigerator/freezers will operate with a colder freezer section (usually between 0° and 5° Fahrenheit) than a conventional unit.

The term *natural convection*, as it is used here, refers not to the movement of heat through a fluid as defined in Chapter Six, but rather to the heat movement created by cold air falling and warm air rising. This is simply the movement of heat because of a temperature differential.

You'll note that the basic refrigeration system components (referred to by most manufacturers as the *sealed system*) are shown in Figure 7-1. The compressor is identified as component #1. Component #2 is the condenser. This type of condenser, known as a *static condenser*, is mounted on the rear of the refrigerator cabinet. It is called static because there is no forced air movement involved in cooling off the hot refrigerant vapor, only the natural movement of heat from the warmer condenser tubing into the cooler *ambient* air.

The evaporator is identified as component #3 and also as component #4 because there was a slight design change during production of this model. Depending on exactly when it was manufactured, the unit will have the part number that corresponds with either identification #3 or identification #4. The manufacturer's method of identifying the numbered components is shown in Figure 7-2.

You'll note that component #5 is identified as being a heat exchanger. This is a standard method of construction used in refrigerator/freezer sealed systems where the capillary tube is attached to the suction line. During construction, the two components are positioned side by side and soldered together so that there will be good metal-to-metal contact between them. Figure 7-3 shows that, while the two components are soldered together, they are still independent of each other in regard to refrigerant flow.

The use of a heat exchanger doesn't change the method of operation of the basic refrigeration system; it only allows the system to work more efficiently. The heat exchanger accomplishes this in two ways: First, since the capillary tube is located after the condenser, it is going to be warm. When it is soldered to the suction line, it is going

CONVENTIONAL REFRIGERATOR

REF. NO.	DESCRIPTION	PART NUMBER	QUANTITY USED	
			C1062 B/M 6W41A or B	C1262 B/M 6W42A or B
	BILL OF MATERIAL INFORMATION			
	C1062 B/M 6W41A AND 6W41B			
	C1262 B/M 6W42A AND 6W42B			
	SEALED SYSTEM PARTS			
1	Compressor (Includes Overload, Relay and Drier)	RA869	1	1
2	Condenser-External Mount (Includes Mounting Parts)	RA43612-1	1	1
3	Evaporator (Includes Drier)	RA43610-1	1	-
4	" " "	RA43611-1	-	1
5	Heat Exchanger (Includes Drier)	RA43605-1	1	1
6	Drier	13900	1	1
	ELECTRICAL PARTS			
7	Relay and Overload Combination	58330-1	1	1
8	Power Cord	52917-1	1	1
9	Lead Cord-Door Switch	28297-16	1	1
10	" " -Compressor	52918-1	1	1
11	" " -Door Switch	54653-10	1	1
12	Temperature Control	52881-29	1	1
13	Socket-Cabinet Light	29814-1	1	1
14	Bulb-Cabinet Light	31956	1	1
15	Switch-Cabinet Light	56432-2	1	1
	FRESH FOOD COMPARTMENT PARTS			
16	Dura-Last V™ Food Liner - R.H.	55829-1	1	-
17	" " V™ " " - L.H.	55829-2	1	-
18	" " V™ " " - R.H.	55830-1	-	1
19	" " V™ " " - L.H.	55830-2	-	1
20	Clip for Mounting Food Liner-Side and Top	54576-1	14	14
21	Clip for Mounting Food Liner-Bottom	58083-1	4	4
22	Escutcheon for Temperature Control	52915-1	1	1
23	Knob for Temperature Control	26621-16	1	1
24	Clip for Retaining Control-Capillary Tube	58453-1	3	3
25	Guard for Light Bulb	52902-1	2	2
26	Shelf-Glide Out	29869-26	-	2
27	" " "	29869-25	2	-
28	Stop-Guild Out Shelf	21697-9	4	4
29	Shelf-Crisper	28530-34	-	1
30	" "	28530-35	1	-
31	Pan-Crisper	56448-5	1	-
32	" "	56448-4	-	1
33	Door for Evaporator Compartment	53266-42	-	1
34	" " " "	53266-43	1	-

Figure 7-2. One manufacturer's method for identifying the components shown in illustrations in a service manual.

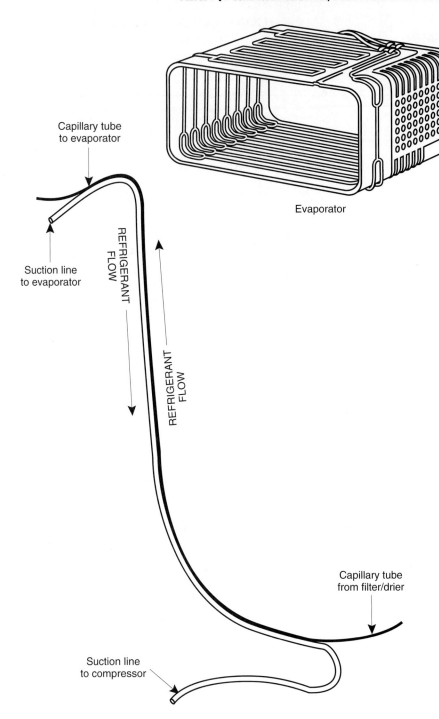

Figure 7-3. A heat exchanger used in a domestic refrigeration system consists of the capillary tube being soldered to the suction line. Note the direction of refrigerant flow.

to give up some of its heat to the cooler tubing. This ensures that the refrigerant will enter the evaporator at a correct temperature and will therefore be able to work as efficiently as possible.

Second, the heat exchanger system allows for more efficient operation through the fact that no liquid refrigerant will get to the compressor. If all the refrigerant is not "boiled off" in the evaporator and some liquid is left to travel down the suction line, it could cause damage to the compressor. You'll recall from Chapter Six that the compressor was identified as being a vapor pump. When we used that term, that's exactly what we meant. A pump designed to compress a vapor can be damaged when it attempts to compress something that can't be compressed—that is, a liquid.

Any liquid that might be in the suction line will vaporize because it now has work to do. The heat provided by the warmer capillary tube provides the warmer temperature necessary for the vaporization process to take place in the suction line.

So the heat exchanger accomplishes two things at once. In the process of heat transfer, the refrigerant enters the evaporator at the proper temperature (something we have to make sure occurs in the first place), and in the process we also make sure that the compressor receives nothing but vapor.

The heat exchanger actually works hand in hand with another method to accomplish the prevention of liquid in the compressor. Refer again to the evaporator in Figure 7-1. You'll note that a section of tubing near the rear of the evaporator appears to be larger in diameter than the rest of the tubing shown in the drawing. This section of tubing is known as an *accumulator*.

An *accumulator* is a storage tank located at the end of the evaporator, and its function is, like the heat exchanger, to aid in the prevention of liquid migration to the compressor.

Component #6 in Figure 7.1 is the *filter/drier*. The function of this component is to filter the refrigerant in the sealed system. Since the filter/drier is located at the end of the condenser, it is filtering the refrigerant as a liquid. This is because of the drier's location at the outlet of the condenser.

(You'll note that there is a piece of tubing connected to the compressor that performs no function. This tube is known as the *process stub*. We'll be discussing its function in Appendix A on sealed system servicing.)

Also, you'll note that, while the inlet of the drier is connected to the condenser, the capillary tube is attached to the outlet of the drier. The capillary tube is shown with its excess length coiled up before it is soldered to the suction line to create the heat exchanger. Coiling the capillary tube is a common practice for most manufacturers because of the design of the sealed system. It may be necessary, for example, to use a total of 15 feet of capillary tube to create the pressure drop necessary for optimum operation of the sealed system. Since the distance from the outlet of the condenser to the inlet of the evaporator may only be 4 feet, coiling the remaining 11 feet of metering device is a convenient method of taking care of the excess tube.

The information to this point doesn't cover everything you see in the Figure 7-1, but it covers everything about the basic refrigeration system. Before we go on to discuss more of the specifics of this illustration, take the time to trace the flow of refrigerant as it leaves the compressor on the discharge line. You'll note that this line is shown as the thick black line that isn't quite connected to the compressor. It is shown this way because of the angle of the illustration. The discharge stub of the compressor can't actually be seen from this angle, so the tube is shown in proximity to the compressor to illustrate that it actually attaches at an unseen location.

As the refrigerant leaves the compressor, it goes to the top of the condenser and, as you can see, travels through the condenser tubing in a back-and-forth pattern and ultimately exits the condenser and flows into the filter/drier. The vertical black lines you see on the condenser are nothing more than steel wires that are soldered to the outer surface of the steel condenser tubing. The wires perform two functions: First, they add strength to the condenser assembly; second, they increase the surface area of the assembly and this allows for more efficient heat transfer (heat rejection into the ambient air).

As the refrigerant leaves the filter/drier, it travels through the capillary tube to the inlet of the evaporator. (You may want to refer again to Figure 7-3 for clarification of how the metering device and suction line are attached to the evaporator.)

After traveling through the evaporator and absorbing heat, the refrigerant is then drawn down the suction line, returning to the compressor, and the cycle is repeated.

Other Components

Now that we're certain of the sealed system parts in the sealed system of a refrigerator/freezer, let's turn our attention to the other items shown in Figure 7-1 and the component listings in Figure 7-2.

At the moment, all the extra components, identifying numbers, and indicating arrows may seem to clutter up the illustration. As you gain experience as an appliance technician, however, you will appreciate the fact that manufacturers use this method in their service manuals to identify each and every component that goes into the construction of a refrigerator/freezer.

As you can see, illustrations of electrical components—such as the motor starting relay and overload protection assembly, clips, covers, grommets, and insulation—are shown and identified. This method of explanation by manufacturers helps the technician to understand how the appliance is constructed and, as a result, allows for a better understanding of how to service the unit.

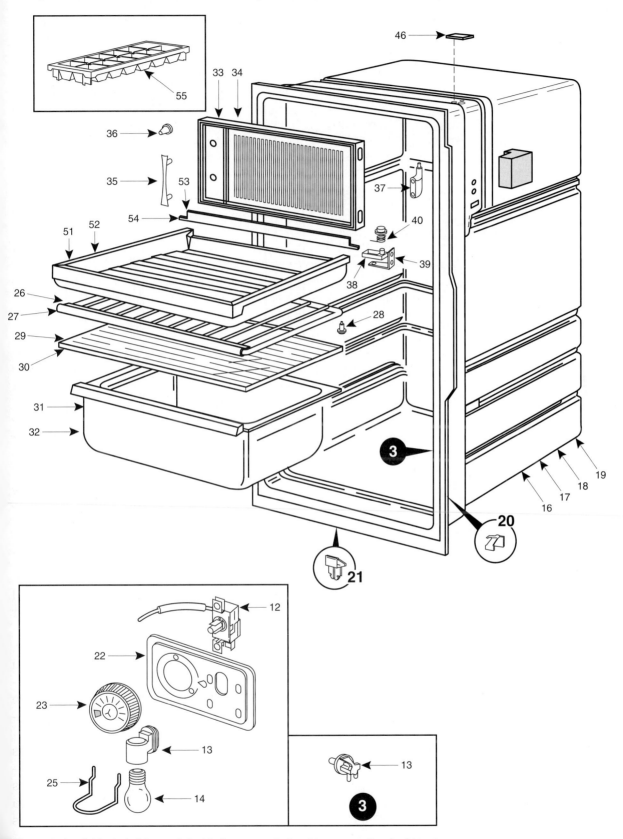

Figure 7-4. Cabinet, shelving, and electrical components found in a conventional refrigerator.

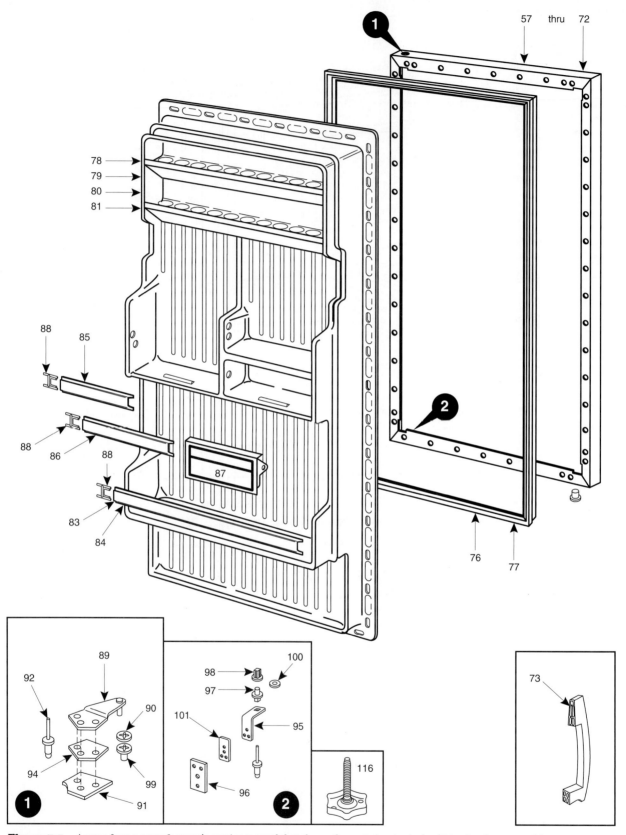

Figure 7-5. A page from a manufacturer's service manual that shows the parts in a typical refrigerator door assembly.

Figure 7-4 is another illustration of a page from the manufacturer's service manual on the same model of conventional refrigerator/freezer. The inner liner, shelving, storage drawers, and some of the unit's electrical components are shown. Studying this illustration will give you an opportunity to understand how the electrical components of the unit, such as the thermostat—component #12 (sometimes referred to as a *cold control* or *temperature control* by some manufacturers)—is fastened to the cabinet assembly. The interior light socket—component #13—and the door switch—component #15—are also shown.

The lesson here is to take the point of view that a service manual gives you an idea of what the customer sees (the knob of the thermostat—component #23—for example). But the manual goes further than that and gives the technician an understanding of what is behind that knob, the body of the control—component #12—and the wiring that connects to its terminals. The sensing bulb of the control is attached to the evaporator section and cycles the compressor on and off according to a cut-in and cut-out temperature of the control. Food items in the freezer don't thaw out because the temperature differential between the cut-in and cut-out temperatures is close enough so that the compressor restarts and maintains the desired temperature.

(For a detailed explanation of the construction and method of operation of electrical components, such as the starting relay, overload protector, and thermostat, refer to Appendix B.)

Aside from the electrical and sealed system sections of a refrigerator/freezer, cabinet components must also be considered when servicing an appliance. Figure 7-5 shows how a manufacturer details things such as the cabinet door, door gasket, hinge assemblies, door handle and cabinet leveling legs.

7.3 CONVENTIONAL REFRIGERATOR ELECTRICAL CIRCUIT SERVICING

As an appliance service technician, you'll find that most of the work performed on a unit will be in electrical component testing and replacement. To perform these tasks in a professional manner with competence and confidence, you must become familiar with wiring diagrams. Manufacturers use two types of diagrams, the schematic and the pictorial, to aid the technician in troubleshooting electrical system problems and replacing components. It's important that you understand the specific intention behind both types of diagrams.

The schematic diagram simplifies the electrical circuits in the appliance, showing them on a *line-by-line* basis. The pictorial diagram shows the physical location of the components and the routing of the wiring harness in a unit. Both diagrams are a help to you when you are solving a problem with an appliance.

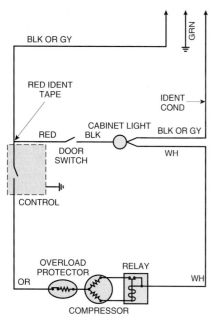

Figure 7-6. A schematic diagram for a conventional refrigerator. This type of diagram makes the electrical circuits in an appliance simple to trace.

Figure 7-6 shows the schematic diagram for a conventional refrigerator and Figure 7-7 the pictorial diagram for the same unit. Take the time now to look both types of diagrams over closely, and keep in mind that these are diagrams from one particular manufacturer's service manual. Another manufacturer may make a similar product that performs the same as this one does, but they may show the electrical circuit in a slightly different manner. Fundamentally, the schematic diagram is a precise method of showing the path of electrical current flow, while the pictorial diagram gives you a better idea of a component's actual location and the way the wires are actually routed within the unit.

As you study the schematic diagram, you'll notice some things that require explanation. We'll begin with the power source and go through a description of what you're looking at in the drawing. As you can see, we've identified the hot wire as the line that goes down the left side of the diagram. The safety ground connection is located in the center, and the neutral wire is shown as traveling down the right side of the drawing. Start with the hot wire and trace the circuit to the first load on the diagram, the cabinet light. You'll note that the manufacturer has identified this hot wire with the abbreviations *BLK or GY*. This is their way of saying that this wire, depending on the date of manufacture, could be either black (BLK) or gray (GY).

As you trace the wire down, you see that it comes to a tie point (junction), which is going to allow for a *parallel* circuit to another load, as well as to the cabinet light. This tie point is just that—a tie point or splice in the wiring system that allows for a connection of the wires for the sake of convenience. No electrical work is done at this point. This could be one terminal on the unit's thermostat, or it could be a *wire nut* that is allowing the connection to take place.

Continue to trace the circuit from the tie point, which in this case is a connection point marked *RED IDENT TAPE* just ahead of the thermostat. You'll note that a red wire now carries power to the door switch that will allow the light to come on when the door is opened. Leaving the door switch, we now have a black wire that actually supplies power to the light socket assembly.

To trace a complete circuit through the light bulb, continue to the right side of the drawing, following the wire marked *BLK or GY* back to the neutral side of the circuit. This wire is also shown as *IDENT CORD*, meaning that, in this unit, the neutral side of the power cord is identified as such. This is sometimes accomplished by marking the wire with a white dot at the point it connects to the unit, or it may be done by showing the difference in the outer insulation of the power cord. One side with ridges, for example, and one side without.

What you've just accomplished is tracing your first electrical circuit from power source (hot wire), through a switch that controls the flow of electricity to the component that does the work, and then back through the neutral wire. This allows for the fundamental thing necessary to get any light to turn on, any motor to run, or any heating element to heat: a complete circuit.

Turn your attention again now to the pictorial diagram shown in Figure 7-7, and trace the same circuit we just discussed. As you can see, even with a circuit as simple as this on a unit as uncomplicated as we can get, it's not as easy to trace from the power source, to a splice, to the door switch, to the light, and back to the neutral side of the power source. Doing this exercise drives home the point that understanding the schematic diagram is essential in servicing appliances. The electrical path is much easier to trace and understand in the schematic. Don't discount the value of the pictorial diagram, though. It's the one that makes those electrical circuits mean something to the person troubleshooting an appliance. Good appliance technicians are able first to identify a schematic symbol, and then "translate" that symbol to what they are actually looking at. The simple switch symbol in the schematic,

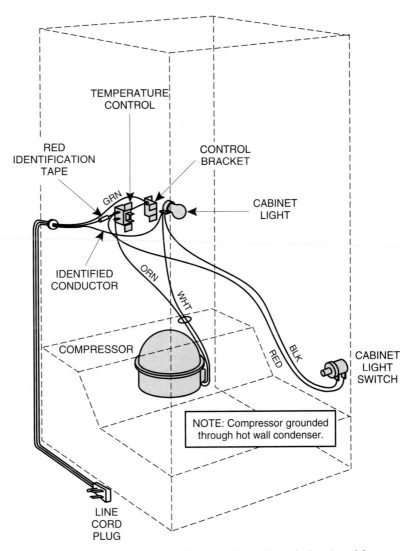

Figure 7-7. The pictorial diagram for an appliance shows the location of the components.

for example, becomes a plunger-type switch with a plastic body and two male spade connections that allows for the attachment of wires.

To continue with our discussion on electrical circuits, refer again to the schematic diagram and begin to trace a circuit from the same starting point as last time. For this circuit, continue past the light switch circuit and instead continue to the shaded area marked *control*. This shaded area represents the thermostat of the unit. (For a detailed explanation of the operating method of a control of this type, refer to the appendix on electrical components).

You'll note that the color of the wire that takes power into the thermostat assembly is a either a black or gray (BLK or GY) wire, and the color of the wire that leaves the thermostat, carrying power down the compressor motor is orange (OR). The orange conductor connects to a device identified as an *overload protector*. This type of device allows electrical energy to pass through it as long as there is no overload condition. It is an externally mounted protection device, and power leaves it through a wire that connects directly to a compressor motor terminal.

To trace a complete circuit, follow the compressor winding that is drawn slanting downward and continue on through the coil of the compressor starting relay. Leave the relay assembly through the white (WH) wire, which is connected to a tie point on the cabinet light assembly where it now becomes a black or gray (BLK or GY) wire. Tracing this wire back to neutral allows you to see the complete circuit that is accomplished through the compressor when the thermostat is closed and calling for cooling. (A complete description of the operation of the relay and overload protector shown in this circuit is offered in Appendix B.)

With the two circuits in this appliance traced, you've accomplished the first step in understanding how to service a unit such as this in the event of a malfunction. To effectively troubleshoot a complaint about the operation of any appliance, you must first understand what a unit is supposed to do. With this accomplished, you are able to use your meter to test for voltage at the appropriate points or to test for *continuity* through a switch.

To illustrate this idea, imagine that there is a complaint in about the unit shown in the diagrams in Figures 7-6 and 7-7. The complaint is that the unit is not cooling, and upon your arrival you are able to assess that the compressor is not running. You understand that, for the compressor to run, 120 VAC must be applied to the overload and relay via the orange and white wire. A test with a voltmeter that shows the voltage applied at this point would indicate that the thermostat was doing its job, allowing the electrical energy to pass, and would therefore be eliminated as a possible cause. You would concentrate on either the relay, overload, or the compressor as the problem.

On the other hand, if no voltage was read at this point, the control could be the defective component. Either it is not reacting to the rise in temperature and not closing to allow the flow of electricity to the overload and relay, or it could be that the contacts inside the control are corroded and not allowing current to pass. In either case, understanding how the thermostat controls the operation of the compressor makes it easier to track down the problem.

7.4 THE CYCLE DEFROST REFRIGERATOR/FREEZER

The *cycle defrost unit*, or *automatic* as it is sometimes called, is similar to the conventional refrigerator in some ways and different in others. It employs the same method of evaporator construction to form the freezer section of the unit, but differs in that there is an extension to the evaporator located in a separate food compartment. This secondary section of the evaporator is sometimes referred to as a *cooling plate* or *humid plate*, or it may be in the form of *serpentine tubing* mounted to the roof of the food compartment.

This unit, rather than using only one cabinet door, will be of the two-door design. Most commonly, the freezer section is located on the top, but it could also be a bottom freezer model or even a side by side model in very rare cases, depending on the age and manufacturer of the appliance. For our purposes, we'll concentrate on the most common cabinet style currently in production, the *top mount* style.

Figure 7-8 shows the system components of a unit of this type. As you can see from this illustration, the freezer section is similar to that of a conventional unit, but a second section of the evaporator is also shown. The condenser (component #8) is also slightly different from the conventional refrigerator. While it does the same job, rejecting (getting rid of) heat picked up by the refrigerant, it is also of the two-piece configuration. The purpose of the two-piece condenser is to create what is known as an *oil cooling loop system* for the compressor, meaning that there are two more tubing connections on the compressor to accommodate this design of operation. (For a detailed explanation of an oil cooling loop system, see the appendix A.)

In this case, the refrigerant leaves the compressor in the same fashion as a standard unit, but then is also rerouted through a closed loop built into the compressor crankcase to accomplish the oil cooling task. In our illustration, the high-pressure vapor exits the compressor via the initial discharge line, which is shown on the left side of the compressor, nearest the bottom. The refrigerant travels into the primary section of the condenser (component #10). It is then rerouted into the closed loop via the discharge return line, which is shown as the tube second from the bottom at the left of the compressor before exiting through the secondary discharge line (third from the bottom at the left of the compressor) to the main condensing section (component #8)

The high-pressure liquid exits the condenser through the filter drier (component #13) and travels along the capillary tube, which is shown as part of the heat exchanger assembly (component #14). As

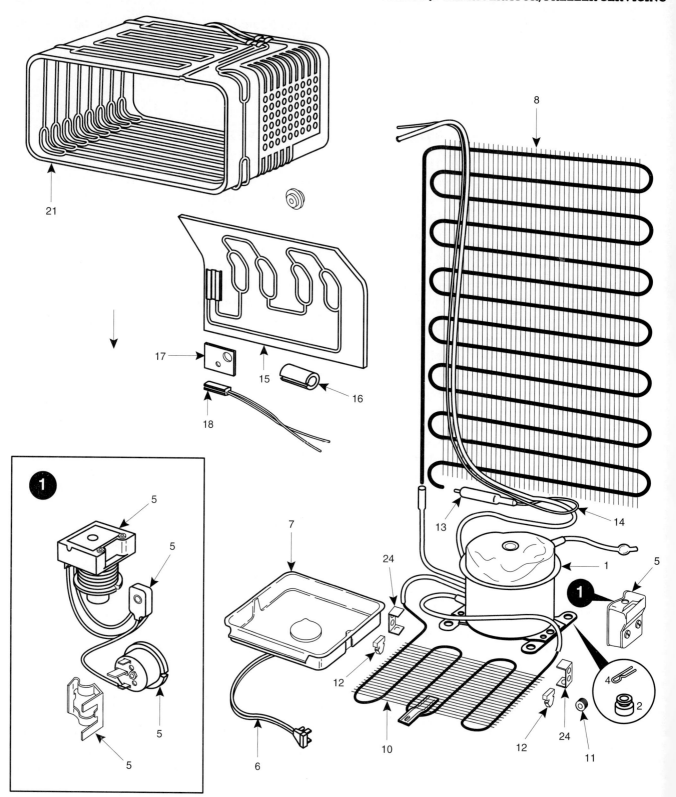

Figure 7-8. The sealed system components of a cycle defrost refrigerator.

with the conventional unit, the refrigerant enters the evaporator via the capillary tube and then exits this section of the refrigeration assembly by way of the suction line, which returns to the compressor (shown in the illustration as the tubing at the top left of the compressor dome).

Other than the miscellaneous clips and grommets shown in the illustration, you'll note that there is a drain pan (component #7). This plastic pan sits atop the primary section of the condenser. Its purpose is to catch the water that runs off the cooling plate every time the unit shuts off and the plate defrosts completely before allowing a restart of the compressor.

In some cases, a trough is located under the cooling plate to catch the water and send it down to the drain pan by way of a hose that is routed down the rear of the cabinet assembly. Other manufacturers may allow the water to drip off the cooling plate and run down the back of the cabinet interior until it exits a small hole in the bottom of the cabinet, which allows drainage into the pan below.

The run cycle of this type of unit is controlled by a thermostat similar to the one that controls the operation of a conventional refrigerator/freezer. The sensing bulb of the control is attached to the cooling plate in the fresh food compartment. When this secondary section of the evaporator reaches the cut-out temperature of the control, the compressor shuts off. The cooling plate, on which a thin coating of frost is built up during the run cycle, then defrosts completely until the cut-in temperature of the control is reached. This restarts the compressor. This on/off cycle is repeated as long as the thermostat is turned to an on position at the chosen setting.

Unlike the conventional refrigerator, the cycle defrost control has a much wider range between the cut-in and cut-out temperatures. This allows for the defrosting of the cooling plate or serpentine tubing that makes up the fresh food section of the evaporator. The fresh food section of the evaporator always defrosts each off cycle, while the freezer compartment must be manually defrosted by turning the unit off for an extended period of time.

In addition to the same basic components found in the illustration of the conventional refrigerator, there are some additional items shown in Figure 7-8. Components #16 and #17 are inconsequential, being only a piece of insulation and a mounting plate. Component #18, however, deserves some explanation. It is a known as a *mold-type heater* (refer to Chapter Four), and it actually performs two functions. First, it prevents the build-up of frost on the tubing that connects the two sections of the evaporator; second, it aids in the prevention of liquid refrigerant causing foaming of the compressor crankcase oil during an off cycle. This function will be easier for you to understand through an explanation of the electrical system, shown in Figure 7-9 (a schematic diagram) and Figure 7-10, (a pictorial diagram).

We'll begin our discussion with a focus on the schematic diagram. As with the conventional refrigerator/freezer already covered, this diagram shows the hot wire on the left and the neutral wire on the right of the drawing. The first circuit to trace will be the divider channel heater wire circuit.

This component is used to prevent sweating on the small section of cabinet that separates the freezer section from the fresh food section. Condensation would occur on this section of cabinet because of its close proximity to the freezer section. Due to the cold temperature in the freezer, the divider channel, sometimes referred to as a *mullion* by some manufacturers, will be cold and consequently will allow moisture in the air to condense on its surface. To prevent this, the manufacturer of this unit provided a low-wattage heater with an adhesive foil back (refer to Chapter Four), which is fastened to the divider channel and as a result warms the metal surface, keeping it above the dew point.

Other names manufacturers use for this component are *mullion heater*, *cabinet drier*, *dew point compensator*, or *center rail drier*. As you can see by tracing the complete circuit from the hot wire to the neutral wire, the divider channel heater wire is energized whenever the unit is plugged in. As long as the unit is working normally, the

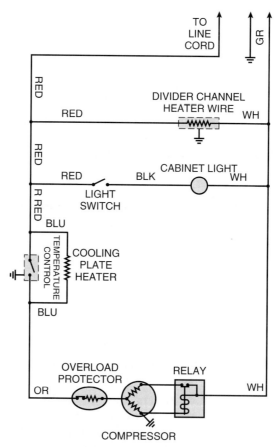

Figure 7-9. The schematic diagram for a cycle defrost refrigerator. This type of unit contains more components than the conventional refrigerator, but reading the diagram line by line simplifies understanding of the appliance.

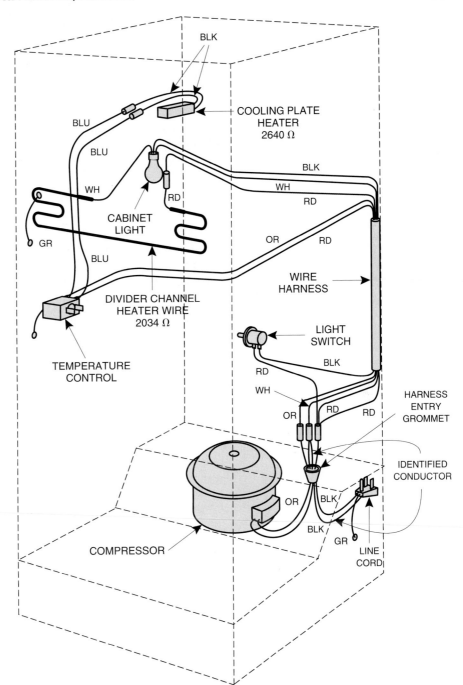

Figure 7-10. A pictorial diagram of the same unit shown in schematic form in Figure 7-9.

mullion doesn't become too hot either during a run cycle or between run cycles. There are times, however, when a customer may become unnecessarily concerned about the temperature of the center rail. If they leave the refrigerator plugged in and turn off the thermostat for defrosting and cleaning, for example, the mullion will become warmer than they recall, and they may in turn call for service because they think there is something wrong with the appliance.

The next circuit on the diagram is the cabinet light circuit. This is a simple circuit that you can trace from the hot wire, through the door switch, then through the light itself to the neutral side of the circuit.

You'll note that the compressor circuit on this unit is the same as the circuit previously discussed on conventional refrigerators. The hot wire passes through the temperature control, through the overload protector, on through the compressor run winding and relay coil, then back to the neutral side of the power source.

While all these circuits are simple for most new technicians to understand, the one that gives those just starting out in the appliance service field the most trouble is the circuit shown as the *cooling plate heater* (sometimes referred to incorrectly by some technicians as a *defrost heater*). This component, as we said previously, has two functions. One is to provide a source of heat to prevent the build-up of an "ice ball" at the point where the cooling plate connects to the evaporator assembly; the other function is to prevent the migration of liquid refrigerant to the compressor during the off cycle. It is wired in *parallel* with the unit temperature control and accomplishes its task through a fundamental law of electricity: *Electricity will follow the path of least resistance.*

The cooling plate heater is fundamentally a resistor in a circuit; therefore, when power is applied, heat will result and voltage drop will also occur. Since the heater is a resistor, when the contacts of the temperature control are closed, there is no circuit through the heater because the path of least resistance is through the temperature "switch." When the temperature control opens, however, the path of least resistance is no longer through the contact points of the control because they are, in fact, separated, but it is through the resistor that allows a complete circuit. The cooling plate heater is designed so that, while it drops most of the voltage in the circuit, it does not use up all the energy. A small amount of current is allowed to continue through the overload protector, through the compressor run winding, and through the coil on the relay; then the connection to neutral allows a complete circuit. This current develops a small amount of voltage drop across the run winding.

What is accomplished with the small amount of voltage allowed to pass through the run winding is that a small amount of heat is generated by the motor winding and, as a result, any liquid refrigerant that finds its way into the compressor crankcase will be boiled off and not cause oil foaming. The small amount of voltage applied across the run winding is not enough to cause the motor to attempt to run or to cause damage to the winding, only to create the small amount of heat needed to solve the refrigerant migration problem. When the temperature in the cabinet rises during the off cycle of the compressor, the temperature control senses it and the control then closes, creating a path for full voltage to be applied to the compressor. The cooling plate heater is deenergized until the control circuit to the compressor is broken again.

7.5   THE FROST-FREE REFRIGERATOR/FREEZER

Sometimes referred to as *forced-air refrigerators*, this type of unit is by far the most popular with consumers since it is self-defrosting and, unlike a conventional or cycle defrost unit, does not have to be shut down on a regular basis to allow for defrosting of the freezer section.

The frost-free unit operates on many of the same basic principles as the conventional and cycle defrost units, while employing a different style of evaporator, known as a *finned type of evaporator*, usually a fan-cooled condenser rather than a static condenser, and the addition of components such as the defrost timer, defrost termination thermostat and the defrost heater, to accomplish the automatic defrosting necessary for the efficient operation of the unit.

Figure 7-11 shows an example of this type of refrigeration system. The finned evaporator, which allows for a forced air operation is shown as component #29. Other electrical components necessary for the operation of a frost-free system, such as the defrost heater (component #30), the defrost termination thermostat (component #33), and the defrost timer (component #25), are also shown.

A defrost system is necessary because of moisture in the air. As the frost-free system operates, it draws air past the tubing that makes up the evaporator section. As the air is cooled by the action of the refrigerant's absorbing heat, the moisture-holding capacity of the air is also reduced, and as a result the moisture condenses on the surface of the evaporator tubing. Since the surface temperature of the evaporator tubing is usually below 0° Fahrenheit, this moisture immediately turns to frost and clings to the evaporator tubing. A build-up of this frost would reduce the efficiency of the operation of the unit because the frost would reduce the air flow through the evaporator and also act as a effective insulator to prevent good contact between the circulated air and the evaporator tubing.

This illustration also shows a different type of condenser often found on frost-free refrigerators. Component #10 is used in conjunction with a cooling fan. This condenser, sometimes referred to as a *jelly roll* style of condenser, is not the only type of fan-cooled condenser you'll see when servicing frost-free refrigerators. Some may be of a different configuration but they are the same in concept: Rather than allowing the refrigerant to be cooled off by natural convection like that of a static condenser, this system uses a fan motor (component #11) to increase the volume of air across the tubing, thereby resulting in a more efficient operation.

Figure 7-12 also shows the finned type of evaporator, but this time used in conjunction with a static condenser. This type of system is not as common as the fan-cooled condenser system. In this illustra-

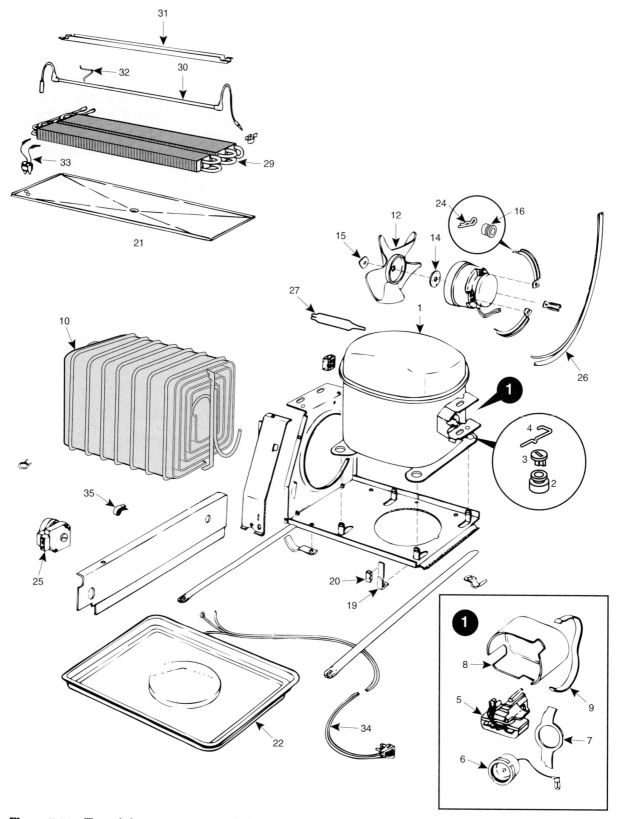

Figure 7-11. The sealed system components of a frost-free refrigerator/freezer. Both a forced air evaporator and condenser are used in this model.

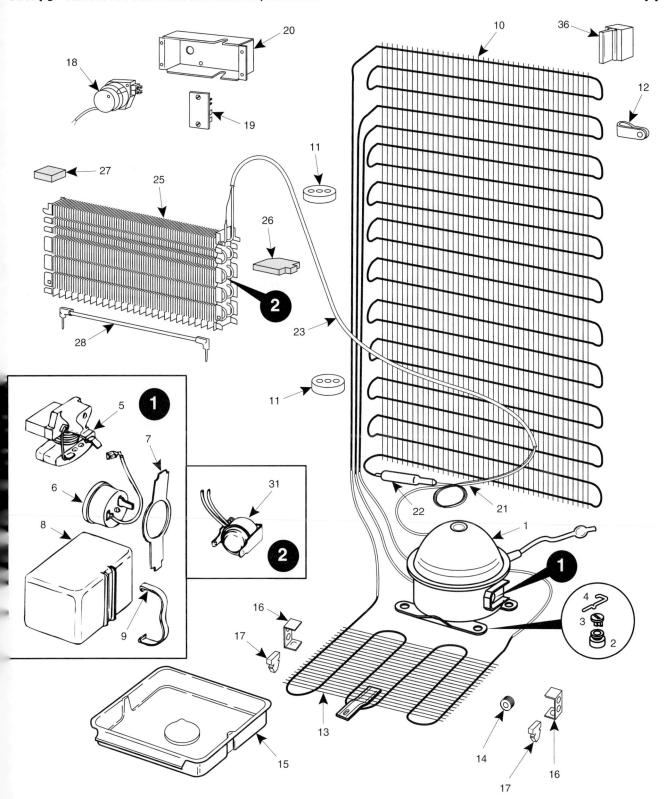

Figure 7-12. A forced air evaporator and a static condenser system of a frost free refrigerator/freezer. This unit uses an oil cooling loop system in the compressor.

tion we get another look at a page from a manufacturer's service manual. In addition to the components you're already familiar with, it also shows some variations in the method of construction of a frost-free refrigerator. Besides the miscellaneous clamps, bumpers and tubing supports (components #11, #12, #16, #17, and #36), it shows the strategically placed pieces of styrofoam (components #26 and 27) that make sure air is forced through the evaporator tubing, as well as an electrical terminal board that is used as a tie point for the wiring harness (#19), and the box mounted to the rear of the cabinet and containing the defrost timer (component #20).

When the cabinet design is top mount (freezer on top), the evaporator coil may be located behind a panel at the rear of the freezer compartment, or it may be located in a horizontal position, mounted under a panel that makes up the floor of the freezer section. In either case, a styrofoam separator insulates the freezer compartment from the fresh food compartment. In the case of a bottom freezer design, the finned evaporator will be mounted behind a panel at the rear of the freezer compartment.

Another design of the finned evaporator can be seen in Figure 7-13, a side-by-side refrigerator freezer. This evaporator is shown along with the method of defrosting it when necessary—in this case, two defrost heaters rather than one.

Regardless of whether it's a top mount, bottom freezer, or side-

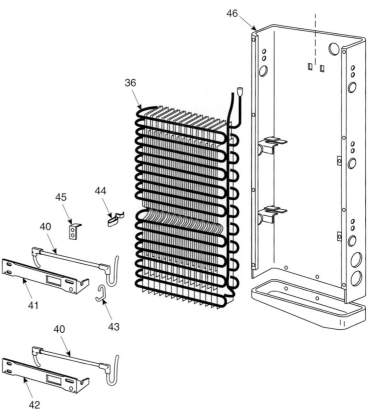

Figure 7-13. A finned evaporator used in a side-by-side refrigerator/freezer.

by-side unit, the fundamental idea to keep in mind about the frost-free refrigerator/freezer is that all the cold air comes from the evaporator located behind or under a panel in the freezer section. You may see separate controls for the freezer and fresh food sections of a frost-free unit, but the fact of the matter is, only one of the controls is electrical; the other is nothing more than an air baffle that allows more or less cold air into the food compartment from the freezer section. Figure 7-14 shows the method of air circulation in a top mount refrigerator, and Figure 7-15 illustrates air flow in a side-by-side unit.

As with the conventional and cycle defrost units, in addition to understanding the refrigeration and air circulation system, an appliance technician must have a complete understanding of the electrical system operation to be able to effectively troubleshoot a problem. Figure 7-16 shows a schematic diagram of a typical frost-free refrigerator/freezer, and Figure 7-17 shows a pictorial diagram for the same unit. While it would be easy to overcomplicate this type of unit, it's not necessary to do so. Just focus on the fact that, in addition to using a modified evaporator and possibly a modified condenser, we've only added the components necessary to accomplish the defrosting of the evaporator. (For a detailed explanation of the electrical components shown in the diagram, refer to Appendix B.)

As you can see from the schematic diagram, the red hot wire (RD) goes into the temperature control, and, when the controls calls for cooling, the blue (BLU) wire carries power to the defrost timer assembly. This hot wire, then, serves two purposes within the defrost timer. First, it applies power to the defrost timer motor to enable it to run and advance toward the defrost mode. Second, it carries power on to a contact point that will allow the energy to leave the timer on

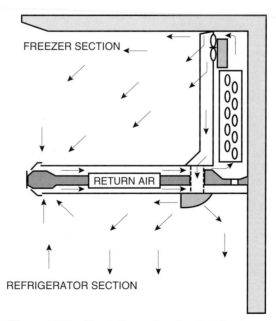

Figure 7-14. The air flow pattern in a frost-free top mount refrigerator/freezer.

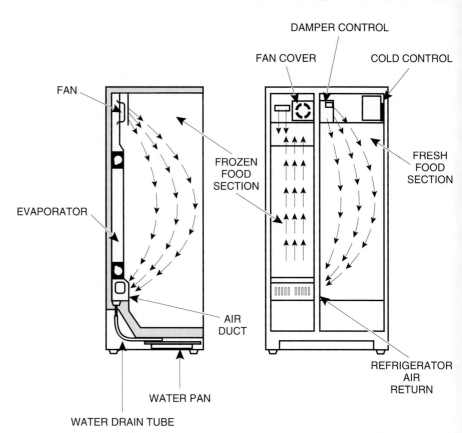

Figure 7-15. The air flow pattern in a frost-free side-by-side refrigerator/freezer.

one of two routes, either to the defrost heater or to the compressor, condenser fan, and evaporator fan.

In this example, the timer is wired in what is known as *cumulative run time*. This means that the timer only advances toward the defrost mode when the refrigeration system is operating, and, after a given amount of "run time" the timer will switch into the defrost mode. The defrost mode, depending on the manufacturer, could range from a low of 18 minutes to a high of 28 minutes.

The schematic in Figure 7-16 shows that the power coming into the timer on terminal one, will leave on terminal two when the cam inside the timer positions the contacts into the defrost mode. The other component in the defrost system is the *defrost thermostat*. This is a bimetal switch that allows power to flow through the resistance heater if the temperature of the freezer section is low enough. The bimetal also senses the rise in temperature and opens, breaking the circuit to the defrost heater. This usually occurs in less than 10 minutes into the defrost cycle. The balance of the defrost cycle time is to allow the moisture to drip off the evaporator and travel either through a tube or down the back of the food compartment, ultimately draining into the drain pan located underneath the refrigerator cabinet.

Once the defrost timer advances out of the defrost mode and

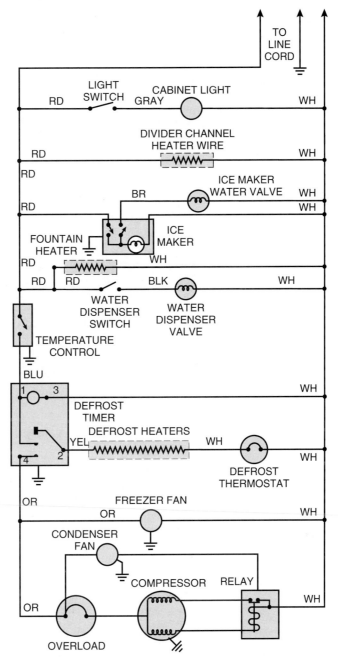

Figure 7-16. A schematic diagram of a frost-free refrigerator/
freezer equipped with an automatic ice maker.

into the run mode, the contacts between 1 and 4 are closed, allowing
for operation of the compressor, freezer fan, and condenser fan. At
the same time these components are energized, the circuit is broken
from 1 to 2. The pictorial diagram in Figure 7-17 is this particular
manufacturer's method of showing the electrical component's physical location.

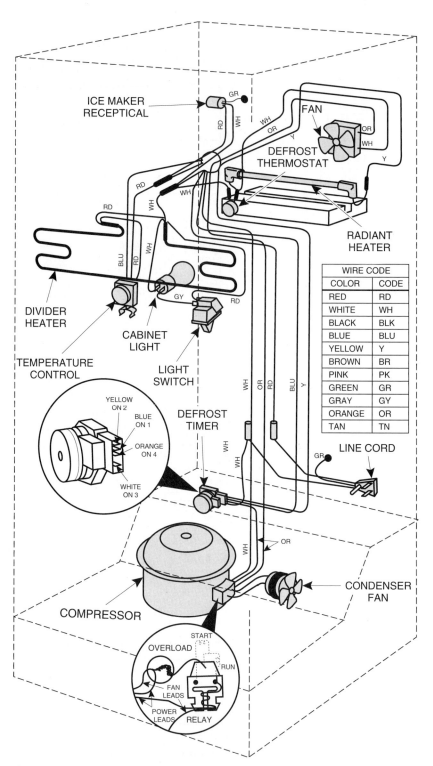

Figure 7-17. A pictorial diagram showing the location of the components of a frost-free refrigerator/freezer.

CHAPTER SEVEN   SUMMARY

The three basic types of refrigerator/freezers serviced by an appliance technician are the conventional, cycle defrost, and frost-free refrigerators. The conventional and cycle defrost units use natural convection to allow for air circulation inside the cabinet. The frost-free unit uses a forced-air system to circulate air from the freezer section to the fresh food compartment.

Conventional and cycle defrost units typically use static condensers, and frost-free units may use fan cooled condensers. One popular configuration of the fan cooled condenser is the *jelly roll* condenser. Other configurations may be used.

A large portion of the appliance technician's work is electrical in nature, and, to effectively service a unit, you need a firm understanding of the wiring diagram. The two types of diagrams technicians use are the schematic and the pictorial. The *schematic* shows the path of electrical flow to the various components within the unit, and the *pictorial* diagram shows the physical location of the parts.

Some specialty electrical components used by refrigerator manufacturers are *dew point compensators*, which are designed to prevent condensation on the refrigerator cabinet and heaters designed to act as resistors. In some cases the manufacturer applies a small amount of voltage to the run winding of the compressor to prevent the migration of liquid refrigerant to the compressor in an off cycle. In this type of unit, a heater, wired in series with the run winding, provides heat at the tail connection of the cooling plate, preventing ice balling at this point.

In the frost-free unit, the defrost timer allows power to the defrost components, which include the defrost heater and defrost thermostat. The heater is a resistance-type element, which may be a heater wire in a glass tube or a Cal-rod-type heater similar to the type of heating element found in an electric range. The *defrost thermostat* is a bimetal switch that reacts to temperature and breaks the circuit to the defrost heater.

Chest and Upright Freezers

LEARNING OBJECTIVES **After studying this Chapter, you will be able to:**

1. Identify the different styles of sealed systems used in food freezers.
2. Trace a simplified wiring diagram for a manual defrost and frost-free food freezer.
3. Differentiate between the air flow patterns in a manual defrost and frost-free food freezer.
4. Explain the door lock system used on domestic freezers.
5. Identify the method used to create a counterbalance system on chest freezer lids.

■ ■ ■

Up to this point, we've discussed sealed refrigeration systems as they apply to refrigerator/freezer combination units. While the same basic principles of heat transfer apply in domestic chest and upright freezers, some methods of design are not commonly found in refrigerator/freezer units, but are more widely used on food freezers. The average appliance technician will, in all probability, not work on freezers as often as refrigerator/freezers.

8.1 TYPES OF FREEZERS

Food freezers and their design are easily understood when categorized like refrigerator/freezers. A *food freezer* is defined as being either a manual defrost model or a frost-free model. *Chest-type freezers*, with the rare exception of one or two models equipped with a supplemental heating element system that is energized by manually setting the cold control in the proper position, are classified as manual defrost units. An *upright freezer* may be either a manual defrost unit or a frost-free unit.

8.2 MANUAL DEFROST SEALED SYSTEMS

Some freezers may come equipped with a refrigeration system design similar to those already discussed, such as a static condenser mounted on the rear of the cabinet or a fan-cooled condenser. In some cases, however, the sealed system design of the manual defrost freezer may appear as the system shown in Figure 8-1. In this case, the condenser is not mounted to the rear of the cabinet, but instead is spot-welded directly to the inner surface of the freezer cabinet.

With this style of design, the entire cabinet surface serves to dissipate heat picked up by the refrigerant as it travels through the evaporator. Another differing feature on this style of sealed system design is the *yoder loop*. As you can see from the drawing, the yoder loop is

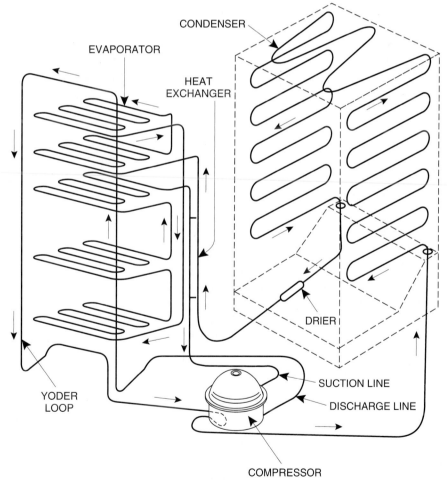

Figure 8-1. Sealed system on a manual defrost upright freezer. The condenser is attached to the inner surface of the cabinet, and the evaporator serves as shelves.

also the oil cooling loop for the compressor. Trace the refrigerant flow from the compressor on the initial discharge line, through the yoder loop, then back to the compressor, and then follow the direction of refrigerant flow to the main section of the condenser that is fastened to the freezer cabinet. This will give you an understanding of the design of the unit.

The yoder loop actually performs two functions. First, it acts as the oil cooling loop (refer to Appendix A for a detailed explanation of the oil cooling loop system), and it also functions as a heat source to prevent the cabinet from sweating. Used in this fashion (as shown in more detail in Figure 8-2), the yoder loop eliminates the need for an electric heater to keep the cold freezer cabinet above the dew point, and it prevent drops of moisture from forming on the cabinet in humid weather. In some cases, rather than refering to this section of condenser tubing as the yoder loop, a manufacturer may identify it as a *perimeter hot tube*. In any case, the functions of the tube are the same.

When you refer again to Figure 8-1, you can continue tracing the sealed system by following the refrigerant flow through the drier, through the capillary tube (noted as a component of the heat exchanger), and through the shelving sections that make up the evaporator. To complete the flow of refrigerant through the entire sealed system, trace along the suction line to the compressor. You'll be able to see that the flow circuit is ready to repeat itself through the discharge line and into the yoder loop.

The evaporator design in this type of freezer may vary. In some cases, the manufacturer may use steel tubing with several passes of steel wire welded to the tubing to make up a shelf. In other cases, the manufacturer may use aluminum tubing and attach it to a lightweight shelf assembly. When the evaporator tubing is attached to the shelves, they are known as *live shelves*. A live shelf is one that not only supports packages, but also carries refrigerant through the tubing. A freezer may also use shelves that are not "live" in conjunction with those that are.

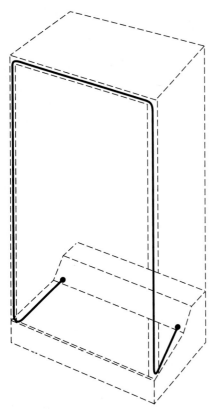

Figure 8-2. Routing of the yoder loop on an upright freezer. Sometimes referred to as a perimeter hot tube, the heat exchange from this section of the condenser tubing prevents moisture from forming on the cabinet.

8.3   CHEST FREEZERS

Another example of the manual defrost freezer is the chest freezer. This type of unit is the most efficient due to the fact that cold air does not escape and warm air doesn't migrate in when the door is opened. Figures 8-3 and 8-4 show the sealed system as it applies to chest freezers. Figure 8-3 shows the steel condenser tubing as it is attached to the inner surface of the freezer cabinet. In this case, it is referred to as the *high side system* because the condenser is the high pressure side of the refrigeration system. Figure 8-4 shows the evaporator tubing as it is attached to the inner liner of the freezer. The reference to the low side system is in regard to the evaporator being the low-pressure side of the refrigeration system.

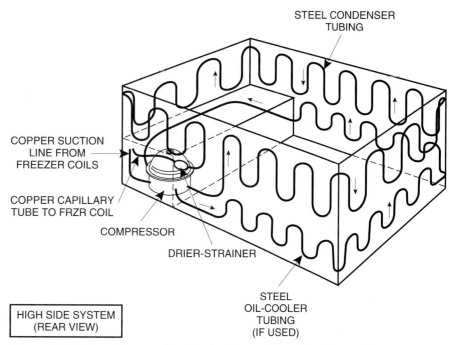

Figure 8-3. The condenser section of a chest freezer. The entire cabinet surface acts to reject the heat from the condenser tubing.

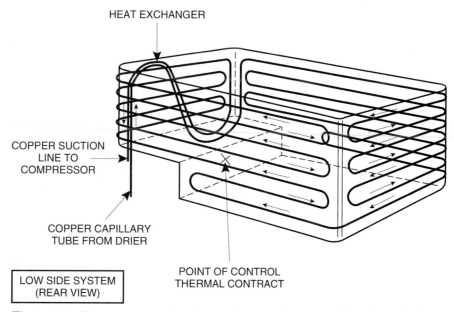

Figure 8-4. The evaporator section of a chest freezer sealed system. Referred to as the low side system since it is on the low-pressure side of the refrigeration system.

The two sides of the refrigeration system are separated from each other by the foam insulation sandwiched between the inner liner and the outer cabinet. The tubing is attached to each respective cabinet section in such a way that the best heat exchange possible is accomplished. This is shown in Figure 8-5.

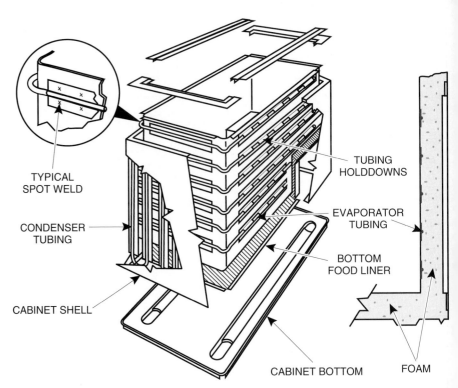

TYPICAL
SPOT WELD

CONDENSER
TUBING

CABINET SHELL

TUBING
HOLDDOWNS

EVAPORATOR
TUBING

BOTTOM
FOOD LINER

CABINET BOTTOM

FOAM

Figure 8-5. A cutaway of the inner and outer cabinet sections of a chest freezer. Insulation prevents heat exchange from the cold evaporator tubing and the hot condenser tubing.

8.4 MANUAL DEFROST FREEZER ELECTRICAL CIRCUITS

To round out our discussion on manual defrost freezers, we'll take a look at a common electrical system for both the upright and the chest freezer. As discussed in Chapter Seven, the two types of wiring diagrams to consider when servicing a piece of equipment are the schematic and the pictorial diagram.

Figure 8-6 shows both types of diagrams for an upright manual defrost freezer. In our illustration, there are few components to consider and the diagrams are, to say the least, uncomplicated. This particular unit utilizes a hot leg of power through a temperature control and overload protector to common on the compressor; the neutral leg of power travels through the coil of the current relay to the run winding of the compressor. (For a detailed description of the current relay as a compressor start device, refer to Appendix B.)

Figure 8-7 shows a pictorial diagram for a typical chest freezer. You'll notice that, in this case, in addition to the compressor and temperature control that make up the main electrical components in a manual defrost unit, this diagram shows a signal light and an optional cabinet light.

The signal light on a freezer serves as a reminder to the consumer that power is being applied to the unit. The cabinet light on a chest freezer is usually a combination light/switch unit that uses a

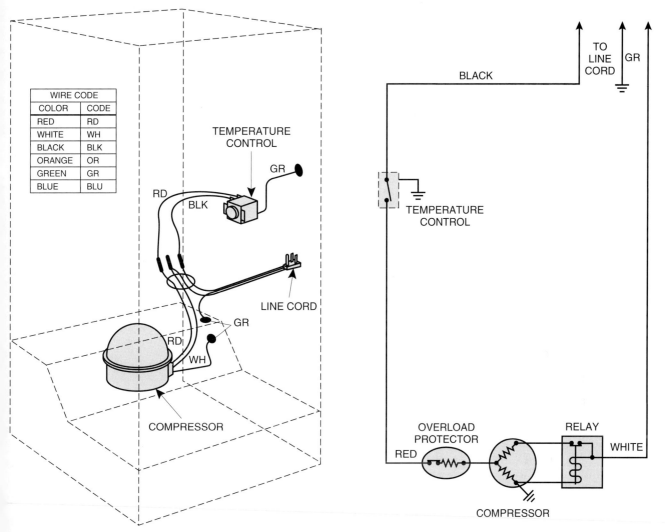

Figure 8-6. The electrical system of a manual defrost freezer is relatively simple, consisting of only a power supply, temperature control, compressor, and start components.

mercury switch allowing the circuit to complete when the lid is raised and breaking the circuit when the lid is closed.

8.5 SEALED SYSTEMS IN FROST-FREE FREEZERS

The sealed system of a frost-free freezer differs from that of the manual defrost unit in evaporator design. Instead of evaporator tubing that makes up shelving and cools the cabinet through convection, the frost-free freezer utilizes a finned evaporator and a forced-air system to accomplish the cooling process. As in manual defrost units, though, the types of condensers may vary from the usual static condenser mounted to the rear of the cabinet, or a fan-cooled unit may be used. Also, in some cases, the frost-free freezer system may be as shown in Figure 8-8 with a condenser and yoder loop system.

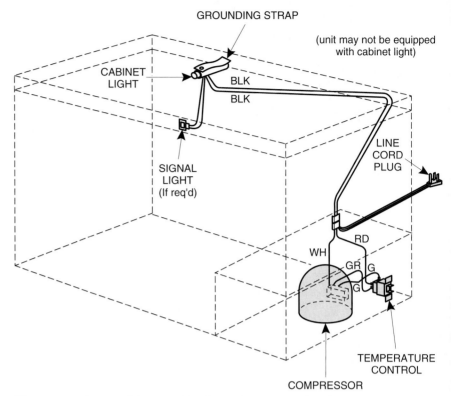

Figure 8-7. A pictorial diagram of the electrical system of a chest freezer. The signal light and cabinet light may be optional.

As in previous illustrations, you can trace the path of refrigerant flow. In this case, the precooler section of the condenser and the yoder loop combine to make up the oil cooling loop system. The refrigerant leaves the compressor and travels first to the precooler. Then it goes through the yoder loop before completing the oil cooling loop and going on through the main section of the condenser. After the refrigerant passes through the drier, it passes through the capillary tube, which is shown in this case as a portion of the heat exchanger system. (For a review of the function of the heat exchanger system, refer to Chapter Seven.)

To complete the cycle, the refrigerant passes through the evaporator where it vaporizes and absorbs heat, then travels down the suction line back to the compressor, and the cycle is repeated.

As we said, this type of system uses the forced-air method to accomplish the cooling process. This type of air flow system is shown in Figure 8-9. As with the frost-free refrigerator/freezer, the frost-free freezer must use an automatic defrost system to keep the finned evaporator coil clear of frost. Proper heat exchange can only be accomplished if the contact surface of the coil is clear, allowing the refrigerant to absorb heat in the manner we discussed in Chapter Six.

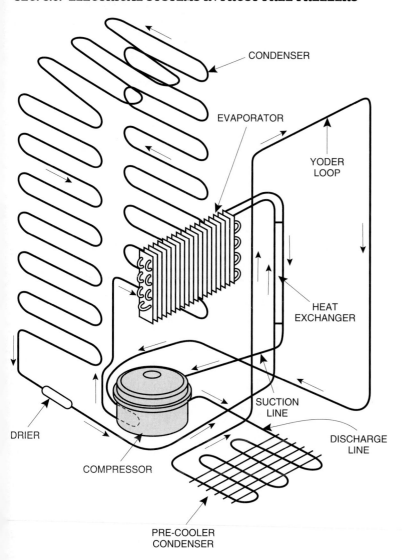

Figure 8-8. Sealed system of a frost-free freezer. The in-wall condenser may be used in conjunction with a finned evaporator.

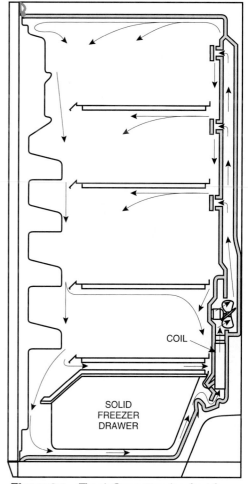

Figure 8-9. The air flow pattern in a frost-free freezer.

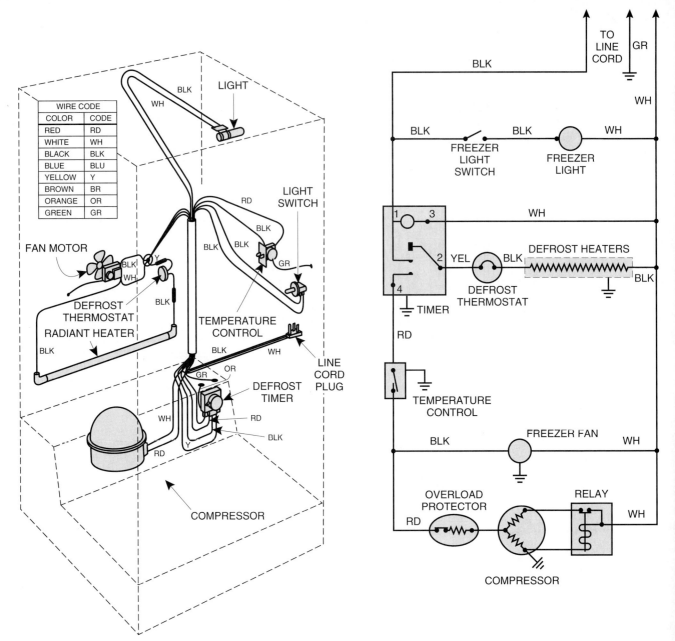

Figure 8-10.   The schematic and pictorial diagram of an upright frost-free freezer. The defrost system in a frost-free freezer works in the same manner as that of a frost-free refrigerator/freezer.

8.6   ELECTRICAL SYSTEMS IN FROST-FREE FREEZERS

The electrical system in a frost-free freezer is more complicated than that of a manual defrost unit due to the fact that more components are required to circulate the air and accomplish the defrosting of the evaporator. Figure 8-10 shows both a schematic and pictorial diagram for a typical frost-free upright freezer.

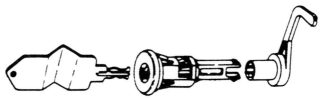

Figure 8-11.   A door lock assembly commonly found on domestic freezers.

The method of operation—a defrost timer, defrost termination thermostat, and heater—is identical to that of a frost-free refrigerator/freezer, such as that discussed in Chapter Seven.

While the pictorial diagram presents the locations of the unit's electrical components, the schematic affords you an opportunity to trace the electrical circuits. You'll note that, in this case, the defrost timer is wired for constant run. Some units may be wired to allow for the advancement of the timer whenever the unit is plugged in. Others may be wired for cumulative run time, which allows for advancement of the defrost timer only when the temperature control is in the closed position and cooling system components are operating.

8.7   DOOR LOCK ASSEMBLIES

One feature found on domestic freezers is the door lock assembly. The door lock, such as the one shown in Figure 8-11 is intended to function as a security device, but its design and location do not hold the door assembly in a totally closed position and is reasonably easy to overcome if necessary. The locking lever of the lock assembly will, on most makes and models, be made of plastic due to UL requirements which state that, should anyone become trapped inside the freezer, the locking lever can be broken or pulled loose from its locked position.

Figure 8-12 further illustrates the method of operation of the freezer locking assembly. Even in the locked position, the door can be pulled open slightly on most models of freezers.

Depending on the manufacturer, a locking assembly may be found on chest-type or upright freezers. Figure 8-13 illustrates a locking assembly in the door of an upright freezer. In the event of a jammed assembly or lost key, the cabinet door could be removed by taking off the top door hinge and lifting the door off the lower door hinge.

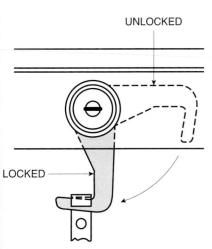

UNLOCKED

LOCKED

Figure 8-12.   The door lock assembly in its unlocked and locked positions.

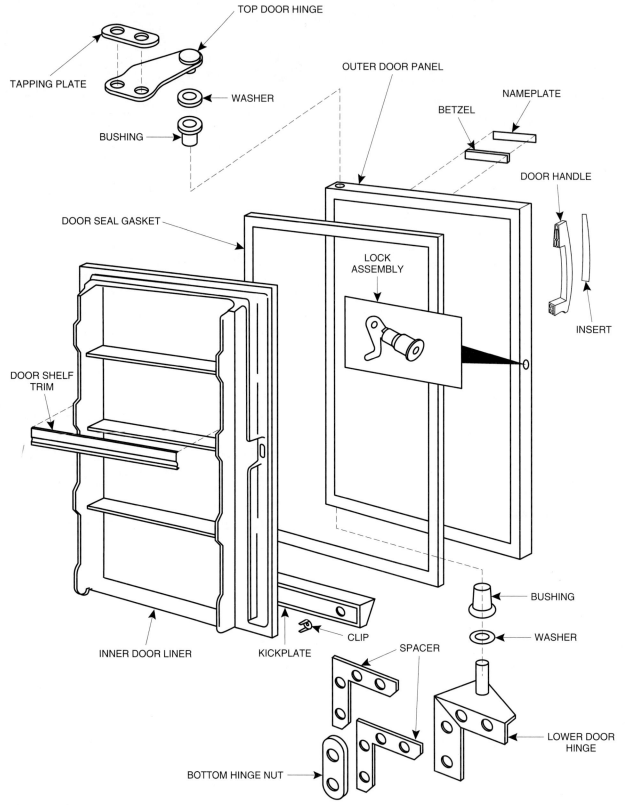

Figure 8-13. An upright freezer door assembly equipped with a door lock. If problems develop with the lock assembly, a technician may be required to remove the door to make the repair.

8.8 CHEST FREEZER HINGE ASSEMBLIES

As an appliance technician, you may occasionally be called on to adjust the lid or hinges on a chest freezer. The chest freezer hinge is designed in such as way as to allow the lid to close tightly onto the chest. This is usually accomplished with a heavy spring inside the hinge itself. The hinge can be adjusted to allow proper fit of the lid onto the chest, and in some cases the spring tension can be adjusted. Figures 8-14 and 8-15 show the location and method of adjustment of a chest freezer hinge.

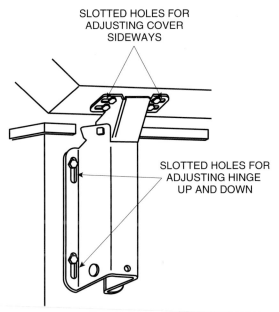

Figure 8-14. A hinge assembly on a chest-type freezer. The adjustments allow for a proper fit of the lid to the cabinet.

Caution! *In the event you find it necessary to remove the hinge assembly on a chest freezer, proceed carefully and maintain pressure on the hinge as you remove the mounting screws. The spring-loaded assembly may snap up and cause an injury if you're not careful.*

CHAPTER EIGHT   SUMMARY

Food freezers may be classified as manual or defrost units, and may be chest or upright in design. A chest freezer will, except in very rare cases, be manual defrost (you may find a chest freezer equipped with electric heater wires that are energized through a manually operated switch), while an upright freezer may be manual defrost or frost-free.

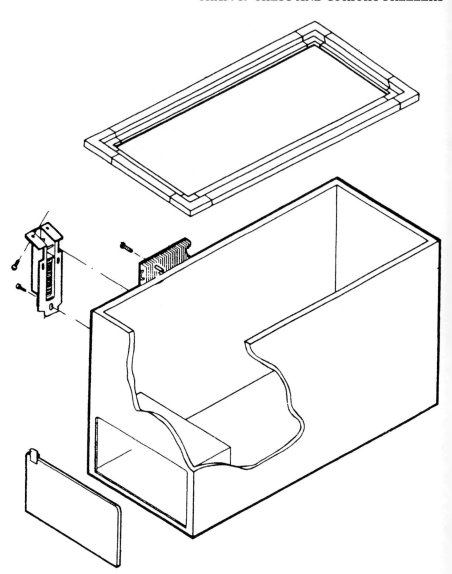

Figure 8-15. The spring-loaded hinge assembly acts as a counterbalance system to ensure that the lid closes tightly. Always exercise caution when removing a spring-loaded hinge assembly.

The sealed system may be similar to that of a refrigerator/freezer with a static or fan-cooled condenser, but it may also be an in-wall condenser system, in which the condenser is attached to the inner surface of the cabinet. It is common to find a yoder loop system in an upright freezer. In some cases, the shelves of the manual defrost freezer make up the evaporator section, while the frost-free freezer uses a finned evaporator coil.

The electrical system of the manual defrost freezer is extremely simple, consisting of only the temperature control, compressor and start components, and in some cases an optional signal light and interior light. The electrical system of a frost-free freezer contains additional components to accomplish the necessary defrosting of the evaporator. The frost-free unit uses a forced-air system.

Food freezers may be equipped with a locking mechanism that can be overcome in the event of an emergency. Chest freezer hinges may be adjusted to allow proper fit of the lid to the cabinet. A spring assembly inside the hinge allows for "counterbalance" that ensures that the lid fits tightly.

Window Air Conditioners

LEARNING OBJECTIVES

After studying this Chapter, you will be able to:

1. Apply the fundamental principles of refrigeration to a comfort cooling system.
2. Differentiate between the different types of compressors used in window A/C units.
3. Explain the purpose of an accumulator, a desuperheater, and a liquid injection circuit on the refrigeration system of a window A/C.
4. Explain the air handling methods used in the operation of a window A/C.
5. Identify the electrical components used in a window A/C.
6. Trace a simple electrical circuit from a window A/C unit.

■ ■ ■

In this unit we will again be discussing heat transfer methods used in a refrigeration system, but our focus will be on comfort cooling systems, specifically window air conditioning units. As an appliance service technician, you'll be expected to be able to troubleshoot and repair window A/C units. Some will be small enough for one person to handle, while others will require a helper to gain access to the system's components.

Window air conditioners, also sometimes referred to as *room air conditioners*, will usually be found in sizes ranging from 4,000 BTUs up to 15,000 BTUs. Some manufacturers may offer units slightly smaller or larger than this, but most of the appliance service technician's work with window A/Cs will be confined within these limits. Your customers will have (or at least should have) purchased a unit of the correct size to match their needs, that is, to cool the room they want to cool. You may find this type of unit installed in a window or in a wall opening.

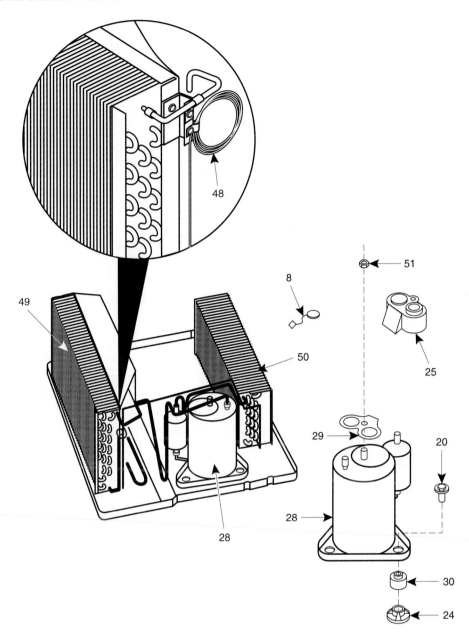

Figure 9-1.   The sealed system components of window A/C unit. As in refrigerators, the capillary tube is used as a metering device. Some larger units may use multiple capillary tubes. *(Courtesy Fedders Corporation).*

9.1   WINDOW AIR CONDITIONER SEALED SYSTEMS

As already discussed, any refrigeration system contains four basic components: compressor, condenser, evaporator, and metering device. Figure 9-1 shows a simplified illustration of a window A/C unit sealed system. In this drawing from a Crosley service manual, the components are easily identified. Component #28 is the compressor, while component #50 is the system's condenser. Component #49 is the evaporator section of the system, and the metering device,

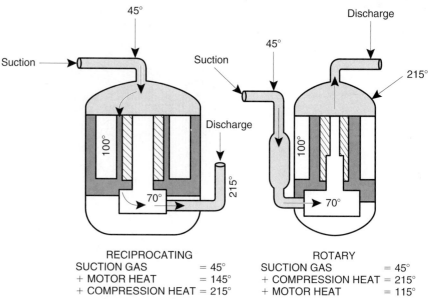

Figure 9-2. Two types of compressors used in room air conditioners: reciprocating (piston) and rotary compressor.

shown as component #48 in the circled area of the drawing, is the system's capillary tube. While the room air conditioner uses the capillary tube as a metering device in the same manner in which the refrigerator/freezer does, the heat exchanger system previously discussed in Chapter Seven is not used in this type of unit. This is only one of the several differences you'll find in the application of a refrigeration system to a comfort cooling system.

The type of refrigerant used in room air conditioners is another of these differences. This is illustrated in Figure 9-2. While reciprocating (piston) or rotary compressors may be used as they are in refrigerators and freezers, the use of R-22 as the refrigerant rather than R-12 creates different operating characteristics. The most noticeable of these differences will be the higher operating pressures and higher discharge temperature. While a refrigerator/freezer will have a discharge temperature of approximately 125 to 150° Fahrenheit, depending on the ambient temperature, a room air conditioner will typically have a discharge temperature of near 215° Fahrenheit. It's important that you keep this factor in mind when servicing window units. The discharge line may be hot enough to burn you. If after servicing a unit you leave electrical wires draped over the discharge line, the heat may damage the wire insulation.

In addition to the difference in operating pressure and temperature between refrigerator/freezers and window A/C units, there are some other differences in sealed system design. One is the use of an accumulator on the suction line. The accumulator is actually a storage tank that is designed to prevent liquid refrigerant from getting to the compressor. You'll recall that we said the refrigeration compressor is a vapor pump, and it is not designed to pump liquid. An accumulator,

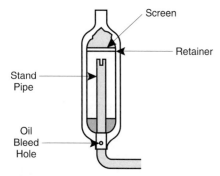

Figure 9-3. A suction accumulator is often found on room air conditioners. An accumulator is a storage tank that ensures that only vapor refrigerant will enter the compressor.

Desuperheating Discharge Gas

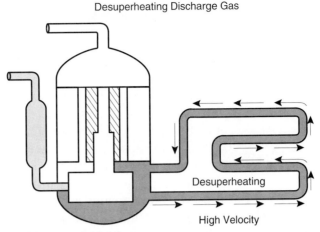

Figure 9-4. When a rotary compressor is used in a window A/C unit, some modifications are made to the refrigeration system. One modification is the use of a desuperheater.

such as the one shown in Figure 9-3, allows liquid refrigerant to enter its chamber, but the standpipe inside the accumulator ensures that only vapor refrigerant is drawn into the compressor via the suction line. Because of the design of a rotary compressor, an accumulator must be used, while in the case of a piston compressor, the accumulator is an accessory which may or may not be found on the system.

In a refrigerator, liquid refrigerant migration is prevented through the use of the heat exchanger system in which the capillary tube is soldered to the suction line, while in room air conditioners, the accumulator performs the same function through a different method.

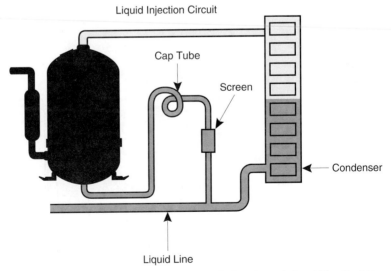

Figure 9-5. Another modification commonly found on window A/C unit refrigeration systems is the liquid injection circuit.

9.2   THE ROTARY COMPRESSOR IN ROOM AIR CONDITIONERS

Manufacturers frequently use rotary compressors in their design of room air conditioners. In the process, they make modifications to the sealed system that are particular to rotary compressor systems. Two of the variables are the addition of a desuperheater and the use of a liquid injection circuit. Both systems are used to help cool the compressor.

The *desuperheater* system is shown in Figure 9-4. The desuperheater is actually a short run of tubing that performs the same function as the condenser, rejecting heat from the refrigerant. It differs from the main condenser section of the unit, however, in that it only cools the initial discharge gas 20° to 30° depending on the design of the unit. It then returns the gas to the compressor in a manner that allows the refrigerant to pick up heat from the motor windings before it exits the compressor via the discharge line at the top of the compressor.

The *liquid injection method* (Figure 9-5), although it performs the same function as the desuperheater, accomplishes cooling of the compressor in a different manner.

When this modification to the sealed system design is used, liquid refrigerant is allowed to flash off after being routed through a capillary tube into the compressor dome near the motor windings. This method is similar in function to the absorption of heat in the evaporator of a refrigeration system. The use of a capillary tube as an expansion device allows the change in state from a liquid to a vapor, and, during this process, heat is absorbed.

These special modifications to a refrigeration system may be used in conjunction with any type of rotary compressor. Two types of rotary compressors are used in room air conditioners and in

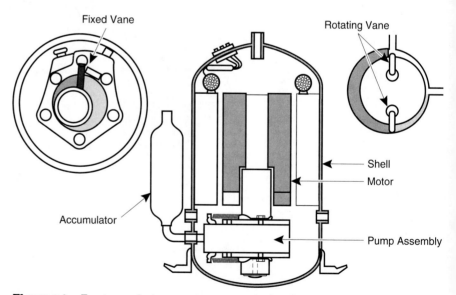

Figure 9-6.   Two types of rotary compressors may be found in room air conditions. The fixed vane and rotating vane.

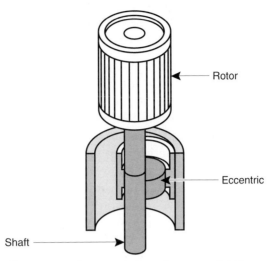

Figure 9-7.   An eccentric is used to accomplish the elliptical motion of the rotary compressor.

other domestic refrigeration equipment: the rotating vane and the fixed vane.

Both designs are shown in Figure 9-6. On the outside, both types of compressors appear the same, and differ from a piston type compressor in that they are more cylindrical in appearance. While the method of accomplishing compression of the refrigerant varies somewhat, both the fixed vane and rotating vane rotary compressors use what is known as an *eccentric* to achieve compression. The term eccentric, such as the one shown in Figure 9-7, refers to a portion of the rotor assembly that is off center and that rotates in an elliptical motion rather than a perfect circle. Figure 9-8 illustrates this method of operation from the beginning of one rotation of the elliptical to its end.

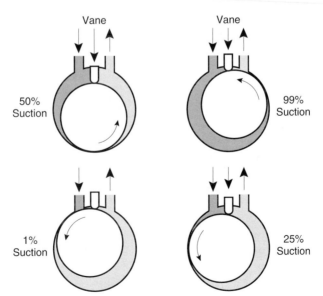

Figure 9-8.   One complete rotation of the rotating vane compressor. *(Courtesy Fedders Corporation)*.

9.3   AIR HANDLING IN WINDOW A/C UNITS

There are two paths of air flow to consider in understanding the operation of room air conditioners: the air over the condenser used to cool the hot refrigerant and cause it to change in state from a vapor to a liquid, and the air over the evaporator coil, which is supplied to the cooled area. Depending on manufacturer's design, the size of the unit, and the model year, you may encounter a wide variety of methods of accomplishing air flow over the evaporator and condenser coils. Figure 9-9 represents only four of the many different designs used by window A/C manufacturers.

With this type of equipment, one fan motor is used to accomplish both patterns of air flow. In most cases, this double shaft fan motor used a *squirrel cage blower* to circulate air over the evaporator and a *propeller-type blade* to create the air flow over the system's condenser. Figure 9-10 shows the positioning of this air handling system, and Figure 9-11 shows in detail how shrouds, separators, and dampers are used to create maximum efficiency in the air flow on both sides of the system.

In Figure 9-11 you can see that the fan motor itself is identified as part #26, the evaporator fan as part #13, and the condenser fan

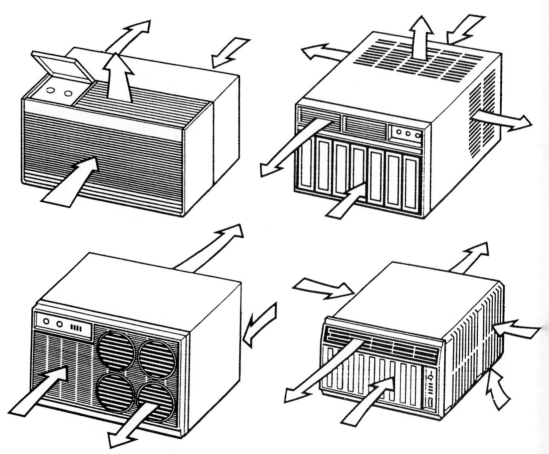

Figure 9-9.   Depending on the design, many different methods of air flow are used in room air conditioners.

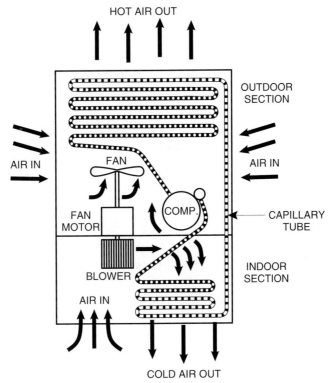

Figure 9-10. A simplified illustration of the air handling system in a window A/C unit.

blade as part #14. A unique factor to note about the condenser fan blade is the ring around the outer perimeter of the blade. This ring, called a *slinger ring* by most manufacturers, has a specific purpose. Condensation takes place on the evaporator coil when the air is cooled. This condensation is allowed to collect in the bottom of the unit, and, as it does so, the slinger ring picks the water up and directs it over the condenser coil. This method of water distribution serves two purposes. First, the condensate water is taken care of and doesn't overflow and drip from the pan. Second, the spraying of this small amount of water over the condenser coil makes the unit's operation more efficient. The water helps to cool the condenser coil.

The last things to consider about air handling in a window A/C unit are the actual cabinet or wrapper, the decorative front, and some adjustable method that will allow a unit to fit into different-sized windows. Figure 9-12, which shows one design used by the Crosley company, illustrates these components.

9.4 ELECTRICAL SYSTEMS

As the size of a room air conditioner varies, so does the applied voltage and the current draw of the unit. And, depending on the current draw and applied voltage, the configuration of the unit's power plug

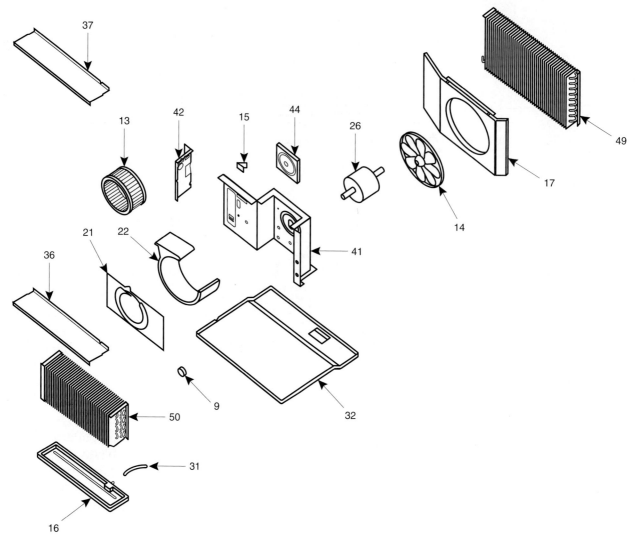

Figure 9-11. A variety of shrouds, separators, and dampers are used to accomplish the efficient operation of the air handling system in a window unit. *(Courtesy Fedders Corporation).*

will also vary. Figure 9-13 shows the different types of receptacles required for various types of window A/Cs. The important thing for you to remember is that each of these receptacles is designed according to the national electric code. Modifying the unit's plug or twisting the blades of the plug to fit into a receptacle can cause damage to the unit and void the manufacturer's warranty.

The only acceptable method of modification is the addition of a three prong adapter on a 120-volt wall plug. In the event you run into an installation in an older home that doesn't have the three-prong plug system, you can use an adapter, providing it is properly installed, such as shown in Figure 9-14. In the event you encounter a modification other than this kind, you should notify your customer of the incorrect application and correct the violation.

Note: *An adapter can only be used if the unit being installed has less than a 7.5-AMP rating. A unit rated higher than 7.5 AMPS*

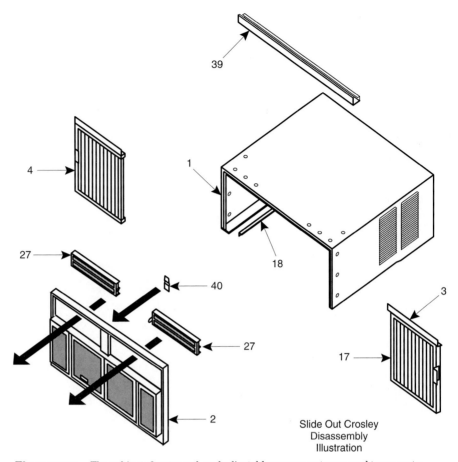

Figure 9-12.   The cabinet, front panel, and adjustable components are used to promote proper air flow in a window unit. *(Courtesy Fedders Corporation).*

would require a separate circuit under manufacturer specifications and the NEC (National Electric Code).

Caution! *The temporary adapter method does not always ensure that the unit is properly grounded. It is extremely dangerous to the consumer and the service technician if a safety grounding system is not established. Always confirm that a proper grounding system exists before using any adapter on an equipment installation!*

The electrical components and wiring harness make up the electrical system of a window A/C unit. While the specific design varies widely from manufacturer to manufacturer, we can cover some

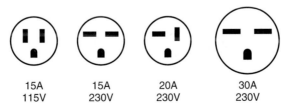

Figure 9-13.   A variety of receptacles are used to accommodate various types of window A/C units.

Adapter Plug (115 Volt Units)

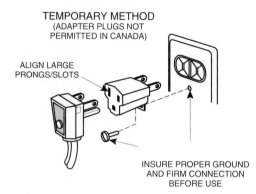

Figure 9-14.   When installing a 115-volt unit, an adapter may be used to allow installation of a three-prong plug into a two-prong receptacle.

fundamental principles in this unit. For additional information on electrical components, see Appendix B.

All units use a main switch to allow the consumer to turn the unit off and on, to select the fan or cooling mode, and to choose the fan speed. Some units have only a high and low speed fan, while

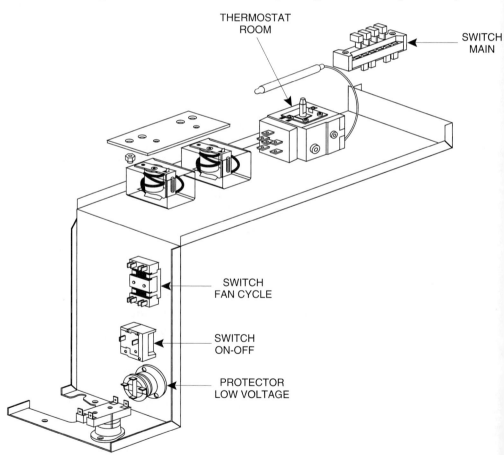

Figure 9-15.   While designs vary, a room air conditioner uses a main switch system and a thermostat to control the operation of the unit.

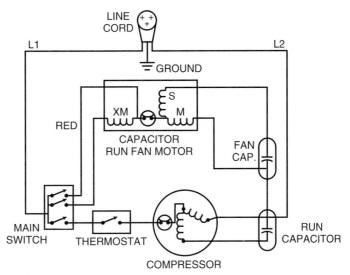

Figure 9-16.   A wiring diagram for a window A/C that uses a capacitor for both the compressor and fan motor.

others may have a three-speed fan. Switching from high cool to low cool doesn't affect the operation of the compressor, only the speed of the fan. When the lower fan speed is selected, the cooling capacity of the unit is reduced proportionately. As much as one-half of the cooling capacity is lost when switching to low cool. Figure 9-15 shows one method of locating the electrical components in a control panel in a room air conditioner.

As with refrigerators and freezers, the electrical system of a window A/C unit can be understood by studying the wiring diagram. Two examples of a window unit wiring diagram are shown in Figures 9-16 and 9-17. One thing you'll notice about the two diagrams is the use of a capacitor in conjunction with the operation of both the compressor and the fan motor. The method of using capacitors varies, however, and you can see this by comparing the two diagrams. In one the capacitor serves only the compressor, while in the other the capacitor is used for both the compressor and the fan motor. In either case, tracing a circuit can be accomplished and helps you to understand the electrical system.

In tracing a circuit in Figure 9-17, for example, we'll assume that the customer has selected the high cool mode on the switch assembly and trace the live circuits. Starting with L1 on the right side of the diagram, we can trace the power leg down to terminal #2 on the main switch assembly. For the fan motor to operate on high speed, the power must leave the switch on terminal #5 and travel via the black wire through the motor winding marked *M* and back to L2 through the tan wire. You'll note that while the tan wire is connected to the run capacitor, the fan motor is unaffected by the capacitor and the terminals are used only as a tie point to make the actual wiring of the harness convenient. (If the customer had selected low speed,

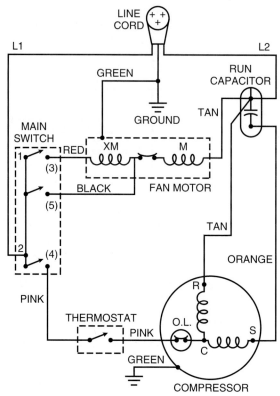

Figure 9-17. A wiring diagram for a unit that uses only a capacitor for the compressor.

the power would leave the main switch on terminal #3 and travel via the red wire through both motor windings to achieve the slower speed.)

The other circuit that must be completed when high cool is selected is the circuit through terminal #4 via the pink wire to the thermostat. In the event the thermostat is closed and calling for cooling, power would then be delivered to the common terminal on the compressor.

The complete circuit on the compressor would actually be accomplished through two paths of flow. The first is through the run winding of the compressor and back to L2 via the tan wire. Again, the capacitor does not affect this circuit and is used only as a tie point. The second circuit would be through the start winding of the compressor via the orange wire. This circuit, as you can see, is wired through the capacitor before completing by returning to L2. This method of wiring allows for the run capacitor to remain in the circuit and thereby cause the compressor to run more efficiently. This is known as a *PSC*, or *permanent split capacitor*, method of wiring a motor to allow for more efficient operation.

CHAPTER NINE   SUMMARY

Window A/C units, sometimes referred to as *room air conditioners*, may vary in size from 4,000 BTUs to 15,000 BTUs, and the operating voltage may be 120 or 240 VAC. The sealed system of a room air conditioner is similar to that of a refrigerator or freezer, using the capillary tube as the metering device but doesn't employ the heat exchanger system. Instead, an *accumulator* is used to prevent the migration of liquid refrigerant to the compressor.

Rotary compressors are frequently used in window A/C units, and some modifications, such as the *desuperheater* or *liquid injection circuit*, are used to help cool the compressor motor windings. The two types of rotary compressors used in window A/C units are fixed vane or rotating vane, and the eccentric is used to accomplish the elliptical motion of the rotary vane.

Air handling is important in window A/C units. The two paths of air flow are over the condenser and over the evaporator. *A squirrel cage blower* is typically used on the evaporator section, and a *propeller-type blade* is commonly found on the condenser side of the unit. A *slinger ring* around the condenser fan blade helps to cool the condenser.

Depending on the size and current draw of the unit, the configuration of the receptacle may vary. Modification of the unit's cord or plug assembly can cause damage to the unit and is in violation of electrical codes. Most window A/C units use the PSC, or permanent split capacitor, method of wiring to ensure efficient operation of the compressor. Some units also use a capacitor for the fan motor.

Automatic Washers

After studying this Chapter, you will be able to:
1. Identify the components that make up the water system of an automatic washer.
2. Differentiate between the various mechanical systems used by manufacturers of automatic washers.
3. Identify the electrical components used in an automatic washer.
4. Trace a simple electrical circuit on an automatic washer schematic.

■ ■ ■

When it comes to servicing laundry equipment, the appliance service technician has to take a somewhat different approach to troubleshooting and repair. When servicing refrigerators, for example, the differences in design are subtle from manufacturer to manufacturer. Once a technician learns the basic skills required, servicing many different makes and models is easily accomplished. But, when you switch from one manufacturer to another when repairing automatic washers, the differences in design of the mechanical and water systems are dramatic, and learning to repair a wide range of models requires a great deal of field experience.

Certain constants can be applied to laundry equipment. Some electrical components, such as water valves and motors, are fairly simple. Once you learn about them while working on one machine, you can apply that knowledge to repairing another type of machine. But there are a lot of different methods of operation employed when it comes to pumps, transmissions, lint filter systems, and even some electrical components such as timers and lid switches.

To help you understand about automatic washers, we'll first offer an overview of the overall method of operation and discuss some of the general differences between one manufacturer and

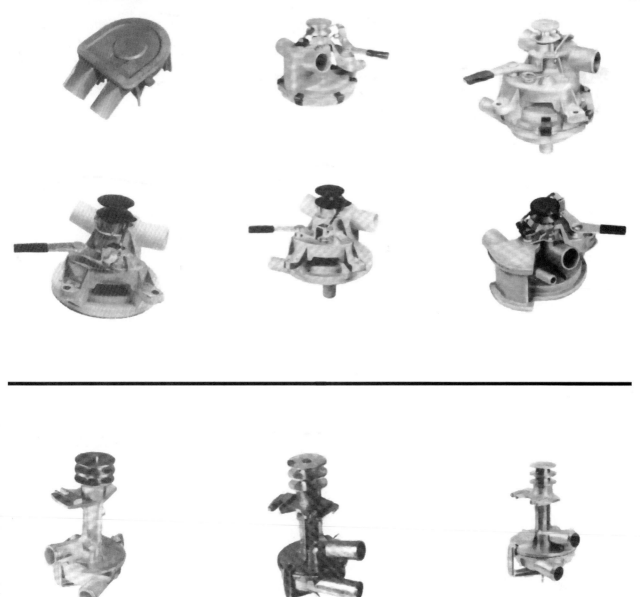

Figure 10-1. Pumps. A representation of different pumps used by two manufacturers, Whirlpool and Speed Queen. *(Courtesy Gem Products, Inc.)*.

another. The first thing you should understand about automatic washers is that they basically perform only four functions: (1) They fill with water, (2) then they agitate, (3) next, the water is pumped out, and (4) the tub spins. The fact that these four functions are performed twice during a complete cycle (one sequence for wash and one sequence for rinse), shouldn't be a cause for confusion or complication. Keeping this thought in mind will help you when you're troubleshooting a problem.

The second thing you need to understand is that, while the four functions that need to be accomplished cannot be changed, the method of getting it done will vary depending on the manufacturer. In some cases, you may encounter a machine in which the four functions are performed by running the motor in one direction for the agitate mode, then reversing direction for the spin mode. (General Electric/Hotpoint, Maytag). In another case, you may find that the motor operates in only one direction and that the switch from one mode to another is accomplished by another method such as shifting inside the transmission assembly (Whirlpool/Kenmore).

And you should be aware that a task as simple as getting the water pumped out of a machine can be performed differently. One manufacturer may pump all the water out before shifting the transmission into the spin mode (many older, belt-driven Whirlpool/Kenmore machines operate this way), while another machine (GE/Hotpoint, Maytag) may be found to switch immediately into the spin mode and pump the water out while the tub is spinning. Every manufacturer makes service manuals available in which they describe the basic method of operation of their unit. Taking the time to review material such as this can save you time and trouble when tracking down a problem in an automatic washer.

10.1 WATER SYSTEM

The water pump of an automatic washer, in many cases, performs two functions. First, it extracts water from the machine and, second, may be used to recirculate water during the wash cycle. When used in this manner, the pump is part of the lint filtering system. Figure 10-1 shows only a few of the styles of pumps used by manufacturers over the years. In this illustration, only two manufacturers are represented. The six shown in the top half of the illustration represent different pumps used by the Whirlpool corporation in various models of Whirlpool and Sears automatic washers. The bottom half represents four different styles of pumps used in Speed Queen machines. Also shown are some of the pump components available to repair the Speed Queen style of pump. In most cases, the pump of an automatic washer is not repaired, but replaced as a complete unit when it fails.

Many pumps are belt driven, while some may be direct drive with their shaft connected directly to the drive motor shaft. One

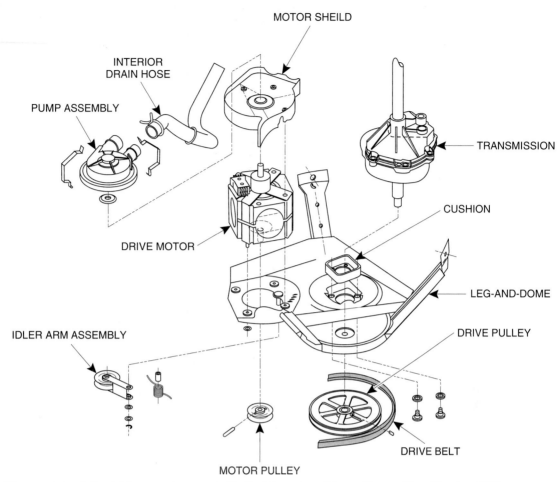

Figure 10-2. A direct drive pump is connected to the drive motor of the washing machine. *(Courtesy WCI).*

Figure 10-3. A lint filter is used on some models of washing machines. Some are self-cleaning; some can be disassembled for cleaning. In some cases, the self-cleaning filter must be replaced when it becomes clogged with lint. *(Courtesy Gem Products, Inc.).*

manufacturer (General Electric) uses a flexible coupler from the motor pulley to the pump pulley to drive the pump. An example of a direct drive pump is shown in Figure 10-2.

When a pump is used as part of a lint filtering system, the change from the filtering mode to the pump-out mode can be accomplished in one of two ways. Either the direction of the pump impeller is reversed to route the water to the drain hose rather than to the recirculating hose, or some type of shift lever on the pump changes the direction of the water flow through the pump body.

In some cases the pump can be disassembled in the event the impeller is jammed with a foreign object such as a sock. Most newer pumps, however, are one-piece molded plastic body units that cannot be disassembled.

10.2 LINT FILTERS

Many different types of lint filters are used on automatic washers. They may range from a self-cleaning assembly that is used in conjunction

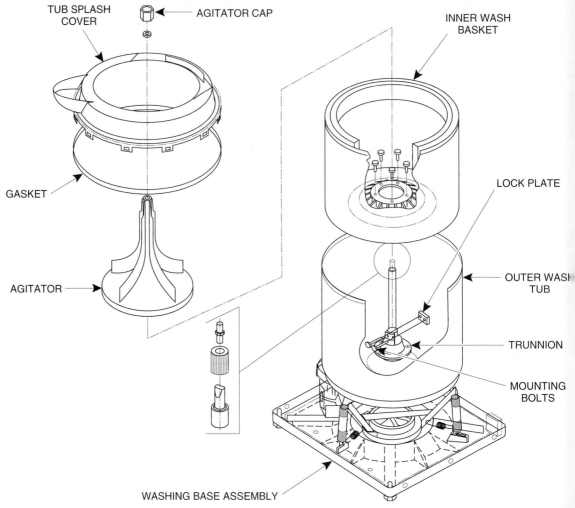

Figure 10-4. The inner and outer tub of a washing machine. Tubs may be plastic, metal, or porcelain on metal. *(Courtesy WCI).*

with a check valve, to a manually cleaned assembly that the customer is instructed to remove and clean at the beginning of each machine use. As with pumps, some filter assemblies are made to be disassembled and cleaned, while others are sealed and must be replaced when they become clogged with lint. An example of a lint filter that can be cleaned is shown in Figure 10-3. Filters that cannot be disassembled are similar in appearance with two hose connections on their body.

10.3   TUB ASSEMBLIES

An automatic washer uses an outer tub assembly to contain the water during the wash and rinse cycles, and an inner perforated tub to hold the clothes. The inner tub is often referred to as the *spin basket.* Depending on the design and manufacturer, either tub assembly may be metal or plastic. Since the transmission assembly extends upward through the center of the tub assembly, a seal is used to prevent water from leaking. An inner and outer tub assembly is shown in Figure 10-4.

Different designs in transmissions require different mounting methods and different designs of seal assemblies. A tub seal can be found to be one that fits into a small opening since only the shaft of the transmission extends through the tub opening, or it can be a larger seal assembly that fits around the entire transmission case assembly. Figure 10-5 shows an outer tub seal found on one type of Whirlpool machine, while Figure 10-6 shows a larger tub seal used in the GE/Hotpoint design of washing machine.

Figure 10-5. A main seal for the outer tub on a Whirlpool washing machine. *(Courtesy Gem Products, Inc.).*

10.4 DRAIN HOSE ASSEMBLY

The drain hose of an automatic washer may be made of rubber or ribbed plastic. On some machines the hose exits the cabinet at a point near the floor, while on others it may exit the cabinet at a point near the top of the machine. Sometimes referred to as a *gooseneck hose*, it is designed to fit into the home drain system standpipe creating a siphon break. If the hose fits too tightly in the standpipe, siphoning may occur during the machine's second fill of a cycle. A drain hose may also contain a check valve assembly. The purpose of a check valve on a drain hose is to reduce air intake by the pump during the agitate cycle. This cuts down on sudsing during agitation. Figure 10-7 shows a drain hose with check valve assembly and a hose with a wire retainer that keeps the assembly in the gooseneck shape.

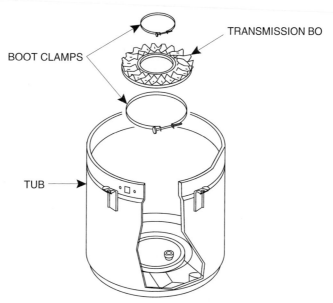

BOOT CLAMPS

TRANSMISSION BO

TUB

Figure 10-6. A main seal used in a General Electric/Hotpoint washing machine.

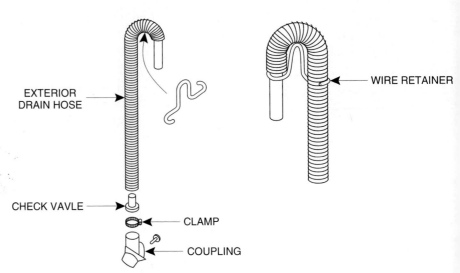

EXTERIOR
DRAIN HOSE

WIRE RETAINER

CHECK VAVLE

CLAMP

COUPLING

Figure 10-7.   The drain hose of a washing machine uses its gooseneck shape to fit into the standpipe of a drain system. *(Courtesy WCI).*

10.5   WATER FILL ASSEMBLY

The water fill assembly of a washing machine consists of the fill hoses, water inlet valve, hoses that allow the water to reach the tub assembly, and a siphon break assembly, sometimes referred to as a *water inlet*. This siphon break (sometimes referred to as an *air gap*) assembly is intended to prevent the siphoning of wash water back into the water supply system in the event of water pressure loss. The water inlet valve, while being a part of the water system, is also an electrical component since it is a solenoid-operated valve that allows the water to flow into the tub when it is energized by the electrical circuit.

The fill hoses are usually made of rubber, although you may find a braided metal hose in use in some applications. In either case, the fill hose uses a brass connection at each end to allow fastening to the plumbing supply system and to the inlet ports on the water inlet valve. Other hoses in the water fill assembly will be fastened with a variety of clamps. A typical water fill system is shown in Figure 10-8.

10.6   TRANSMISSION

The transmission of an automatic washer is considered to be the main component in the unit's mechanical system. A V-belt system that allows connection to the drive motor is the most common method of operation. Some manufacturers, however, refer to their design as *direct drive*. One example of this is the Whirlpool Corporation's Design 2000 unit, which was phased into production in 1982.

As mentioned at the beginning of this chapter, there are vast differences in the method of construction and operation of transmissions among different manufacturers.

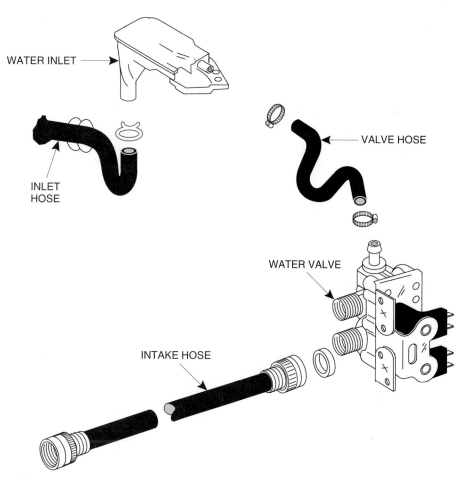

Figure 10-8. A typical water fill system for a washing machine. The water valve is solenoid-operated.

Figure 10-9 shows a Whirlpool/Kenmore belt drive transmission. This type of transmission is known as a *stationary style*. In this case, the motor runs in only one direction. The shifting from agitate to spin is accomplished by energizing solenoids, which pull on cam bars and as a result shift gearing assemblies inside the transmission. The Whirlpool/Kenmore transmission can be disassembled for repair.

Figure 10-10 shows the type of transmission found in a GE/Hotpoint washing machine. This type of transmission is also a stationary type but the method of operation is different. The unit functions in the agitate mode when the motor runs in one direction, and to accomplish the spin mode, the motor reverses direction. With this method, the gearing assembly inside the transmission locks in place and spin is accomplished. In the event of failure of this type of transmission, the standard procedure is to replace the entire unit, buying a rebuilt transmission. Appliance parts dealers who handle this type of transmission price the rebuilt unit on a "with exchange" basis.

Figure 10-11 shows a Maytag transmission. This type of transmission is not a stationary unit, meaning the entire gearcase spins during the spin mode. It is used in conjunction with a brake and clutch assembly.

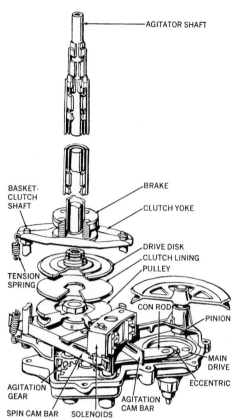

Figure 10-9. An illustration of the belt-driven transmission used in Whirlpool and Kenmore washing machines.

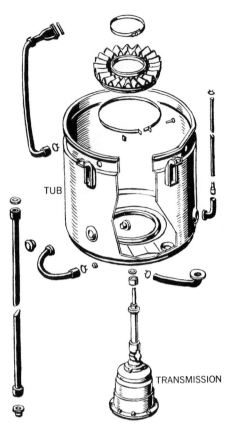

Figure 10-10. The General Electric/Hotpoint transmission is replaced as a complete unit.

Figure 10-12 shows another type of transmission which spins during the spin mode of the washing machine. This unit is the type used in a Gibson washing machine.

As you can see from the wide variety of transmissions, servicing different makes and models of units requires an appreciable amount of study on each individual unit, as well as much field experience with different manufacturer's equipment. We've shown only four different types of transmissions, and there are several more manufacturers of automatic washers.

10.7 DRIVE MOTORS

Most automatic washers will make use of a two-speed motor. A motor speed of 1,725 RPM translates to a "normal" cycle of a washing machine, and a motor speed of 1,140 RPM translates to the "gentle" or "delicate" cycle of a machine. In some cases, you may find yourself repairing a machine that uses only a one-speed motor. This type of unit, referred to as a *bottom-of-the-line* unit, has only a normal cycle of operation. As far as direction of rotation is concerned, some units may use a motor that runs only in one direction while others may operate in clockwise and counterclockwise directions.

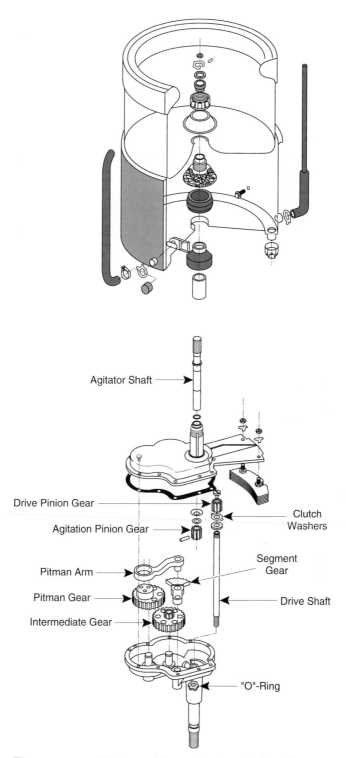

Figure 10-11. A Maytag washing machine transmission. *(Courtesy Maycor, Inc.)*.

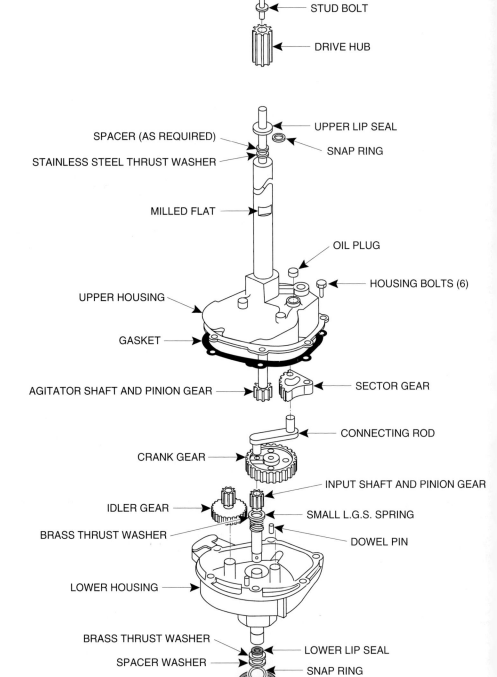

Figure 10-12. The style of transmission used in a Gibson washing machine. *(Courtesy WCI).*

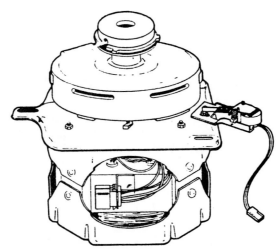

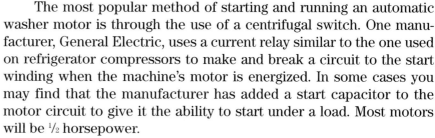

Figure 10-13. A drive motor and clutch assembly found in GE and Hotpoint washing machines.

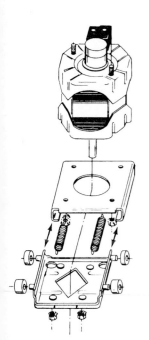

Figure 10-14. A Maytag drive motor and its mounting system. *(Courtesy Maycor).*

The most popular method of starting and running an automatic washer motor is through the use of a centrifugal switch. One manufacturer, General Electric, uses a current relay similar to the one used on refrigerator compressors to make and break a circuit to the start winding when the machine's motor is energized. In some cases you may find that the manufacturer has added a start capacitor to the motor circuit to give it the ability to start under a load. Most motors will be $1/2$ horsepower.

Manufacturers use a variety of methods to mount and operate the motors on their automatic washers. Some type of clutch or slip assembly will be used to allow the spin basket to pick up speed slowly. Figures 10-13 and 10-14 show only two of the several methods of operation used. The GE/Hotpoint motor uses a clutch assembly that fastens directly to the motor shaft. The Maytag design shown in Figure 10-14 is mounted to a motor carriage that uses springs and rollers to accomplish the operation of the motor.

10.8 TIMERS

The *timer* is the one component that is most commonly mis-diagnosed by inexperienced technicians. The reason is that, for many, the timer seems like a complicated component with anywhere from 10 to 20 wires connected to its terminals. The timer, while it controls the entire operation of the machine and will have a large wire harness attached to it, doesn't have to be considered complicated. It's nothing more than a series of switches that are allowed to make or break contact according to the timer's cam and motor. Figure 10-15 shows one type of timer used on automatic washers.

The exact method of construction of a timer varies from one manufacturer to another. Some may use separate terminals to allow

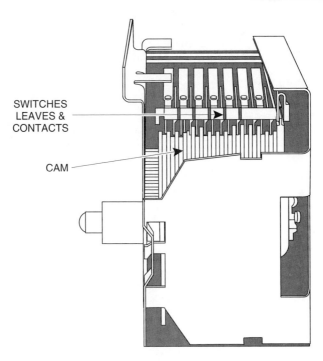

Figure 10-15. A washing machine timer is a cam-operated component that is the heart of the unit's electrical system.

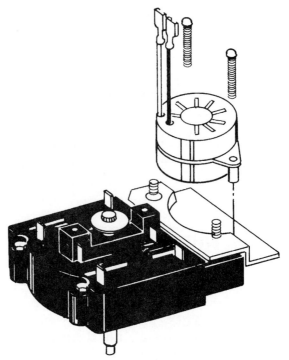

Figure 10-16. In some timers, the motor can be replaced separately. *(Courtesy Maycor)*.

for wiring harness connections, while others may use a quick disconnect harness. These two styles of timers are shown in Figures 10-16 and 10-17. Depending on the design of a timer, the motor may be

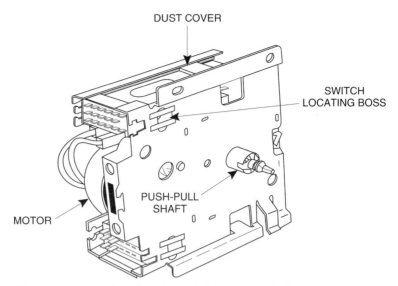

DUST COVER

SWITCH
LOCATING BOSS

PUSH-PULL
SHAFT

MOTOR

Figure 10-17.   A timer that uses a quick disconnect wire harness.

changed without having to replace the entire timer assembly. Other than the motor, most timers are not field repairable and any failure requires that the entire unit be replaced. One exception to this rule is the GE/Hotpoint timer. This timer, known as the *clamshell* timer, can be repaired in the field. Figure 10-18 shows this type of timer.

When replacing a timer, in addition to new units, rebuilt components can be purchased on a cost-plus-exchange basis. Service manuals provide step-by-step instructions for diagnosing and replacing timers.

The timer is located in the console assembly of the automatic washer along with the other components that control the cycles of the unit. The *water level switch* (sometimes referred to as a *pressure switch*), speed selector switch, and water temperature switch allow the customer to select correct amount and temperature of water and desired motor speed. A console assembly with these electrical components is shown in Figure 10-19.

10.9   WATER LEVEL SWITCH

The water level switch allows the customer to select the correct water level for the size of the wash load. These switches are manufactured in many different configurations. Some are push button, some are rotary, and some use a sliding lever. No matter what the style, they are all diaphragm-operated switches. That is, they operate on the principle of air pressure overcoming the selected spring pressure and as a result breaking the circuit to the water valve and stopping the flow of water into the machine. The second function this switch performs is to provide a circuit to the drive motor of the washer. It is actually a single pole, double throw (SPDT) switch that breaks one circuit while at the same time making another circuit.

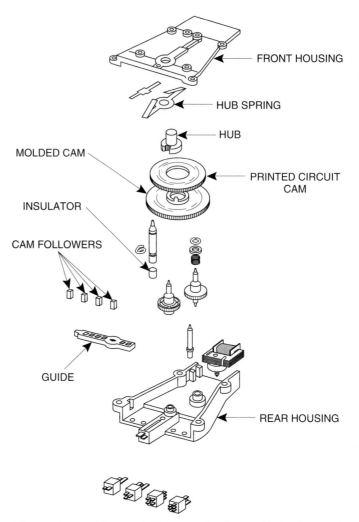

Figure 10-18.   The clamshell timer can be disassembled and repaired.

Figure 10-20 shows the air dome that supplies the pressure to operate the switch. As the level of water rises in the tub, the pressure in the dome increases. This pressure travels along the plastic tube that leads to the switch located in the console and the contacts inside the switch are broken.

We can trace this sequence in Figure 10-21. In this example, a push button switch is used and the control button A has been depressed to select an extra large load. Pushing this button results in locking the tension bar B in a position that exerts maximum pressure on spring C. Set in this position, the tub will have to fill all the way to exert enough pressure in the dome to overcome the spring pressure selected. Pushing a different button would put less tension on the spring, and a smaller amount of water would be required to apply enough pressure to overcome the spring and break the electrical contacts inside the switch.

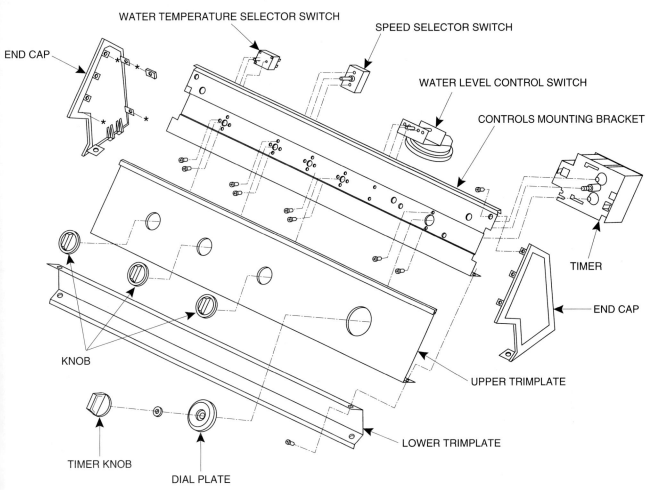

Figure 10-19. A typical console of a washing machine. *(Courtesy WCI).*

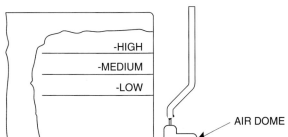

Figure 10-20. An air dome supplies the air pressure necessary to operate a water level control switch.

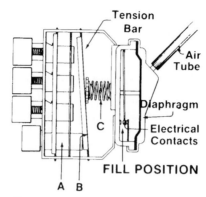

Figure 10-21. A push button water level
control switch. *(Courtesy Maycor).*

The most common use of the pressure switch allows for low,
medium, and high selection, but some switches are infinite and
adjust to any level between the minimum and maximum. An illustration of the three-position rotary switch is shown in Figure 10-22.

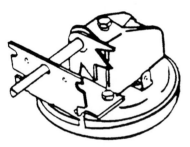

Figure 10-22. A rotary style water level
control switch.

10.10 MOTOR SPEED SELECTOR SWITCH

When the motor speed is not selected through a setting of the timer, a
speed selector switch is used. This switch, like water level control
switches, can be push button or rotary. Its function is to energize
the appropriate motor winding to select either high or low speed for
normal or gentle wash and rinse.

On some makes and models of automatic washers, you can
switch from one speed to another while the machine is in operation
and no damage will occur. With some makes, though, this can cause
trouble. In the case of a GE/Hotpoint washer, for example, changing
the speed selector switch while the unit is in operation will cause
damage to the clutch assembly.

CHAPTER TEN **SUMMARY**

When servicing an automatic washer, the three systems to consider
are the water system, mechanical system, and electrical system. The
water system consists of the pump, lint filtering system if used, drain
system, tub assemblies, and water fill system.

The pump of an automatic washer serves two purposes in most
cases: to circulate the water in the lint filtering system and to expel
the water from the machine. Depending on the manufacturer, pumps
may be direct drive, belt-driven, or coupler-driven. Lint filters and
pumps may, in some cases, be disassembled to clean and reassem-
bled. Many pumps cannot be taken apart because they are one-piece
molded body units. Lint filters may be self-cleaning and may require
replacement if they become clogged with lint.

An automatic washer uses an outer tub and an inner tub, sometimes referred to as a *spin basket*. The tubs may be plastic, metal, or porcelain on metal. A seal must be used to prevent leakage at the opening that surrounds the transmission shaft. An automatic washer drain hose uses its gooseneck shape to fit into the standpipe of the drain system, and some drain hoses may contain a check valve to prevent excessive sudsing during the agitate mode.

The water fill system of an automatic washer consists of the water inlet valve, which is a solenoid-operated valve, the fill hoses, and the siphon break assembly. The siphon break is required if there is a loss of water pressure in the water supply system.

The method of operation of the transmission assembly varies widely from one manufacturer to another. Some transmissions are stationary, while some spin the gearcase during the spin mode.

Drive motors in automatic washers are usually $\frac{1}{2}$ horsepower in capacity. In most cases, they are two-speed motors, operating at 1,725 RPM for normal wash and at 1,140 RPM for gentle cycle. The centrifugal switch is the most popular method of operating the automatic washer motor.

The *timer* is the heart of the electrical system of an automatic washer and is frequently misdiagnosed by inexperienced technicians. In some cases, the motor can be replaced separately, and most manufacturers require that the entire timer be replaced in the event of failure. Some timers have individual connections and some are quick disconnect harness timers.

The *water level control* switch is also referred to as a pressure switch. It is a diaphragm-operated switch that reacts to air pressure, shutting the water off at the selected level and starting the motor on an automatic washer. A motor speed selector switch is used on some automatic washers.

Clothes Dryers

LEARNING OBJECTIVES   **After studying this Chapter, you will be able to:**
1. Identify the electrical and mechanical components of a clothes dryer.
2. Explain the air flow system in a clothes dryer.
3. Identify the components in the ignition and burner system in a clothes dryer.
4. Trace a simple electrical circuit in a clothes dryer.

■ ■ ■

Clothes dryers are simple in their design and construction, and they do not require service as often as other major appliances. Just as automatic washers have different systems that join together and perform a function, the clothes dryer has four different systems: electrical, mechanical, air flow, and heat source. The heat source may be electric heating elements or gas. In the event of gas, today's dryers can be converted from natural gas to LP with a few simple steps. Regardless of the heat source, the mechanical, basic electrical, and air flow systems will be the same.

11.1   ELECTRIC DRYERS

Standard-sized electric dryers operate on 240 volts, and gas dryers require only a 120-volt circuit, plugging into a standard wall outlet. Figure 11-1 shows a typical power cord (often referred to as a *pigtail* in manufacturer's service manuals) and the standard wall receptacle you'll see on a standard size electric dryer. You may see a small apartment-sized electric clothes dryer that operates on a standard 120-volt circuit, but standard-sized dryers that are expected to

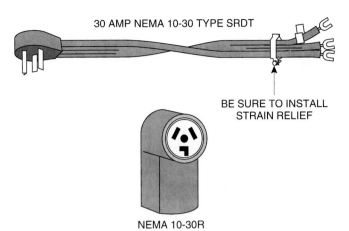

30 AMP NEMA 10-30 TYPE SRDT

BE SURE TO INSTALL
STRAIN RELIEF

NEMA 10-30R

Figure 11-1. An electric dryer requires a 230-volt 30-amp circuit. The power cord of an electric dryer is known as a pigtail, and the receptacle is much different from a standard 115-volt receptacle.

handle a full load from an automatic washer will require the higher-voltage circuit.

In newer homes, the 240-volt circuit will be protected by a two-pole circuit breaker while in older homes, you may find two plug-type fuses used as protection on the electrical circuit. A standard 240-volt electrical circuit for an electric dryer will be rated at 30 amps. This higher-amperage circuit is necessary because of the high current draw of the resistance-type element used in electric dryers. The resistance-type element is the nichrome wire type, often referred to as a *ribbon-type heater*, and is similar in appearance to a heating element in a toaster.

Figures 11-2 and 11-3 show two typical methods of mounting a heating element in a clothes dryer. As you can see, some manufacturers such as General Electric and Maytag, mount the heating element in a plenum located at the rear of the dryer cabinet, while others, such as WCI (Frigidaire, Westinghouse, and Gibson) will mount the heating element in a housing assembly on the base of the dryer cabinet. In other cases, such as Whirlpool (Sears), you may find the heating element in a heater box located behind a cover at the rear of the dryer cabinet. As mentioned, when it comes to laundry equipment, manufacturer's designs differ radically, but the basic performance of the system is the same. In the case of a dryer, air is drawn over a heat source, a motor turns a belt-driven drum to tumble the clothes and operate the blower, and other electrical components control the timing of the cycle and the operating temperature of the unit.

On older dryers, a pulley-to-pulley V-belt drive system was used to accomplish the turning of the drum. Newer units, however, use a much thinner belt that is routed around the entire circumference of the drum. Figure 11-4 shows how the belt is used to tumble the drum. An idler pulley is used to maintain tension on the belt.

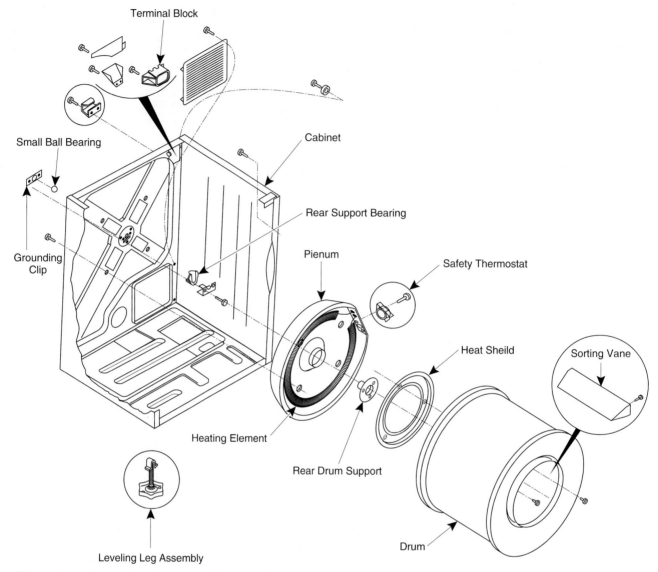

Figure 11-2. One method of mounting the heating element in an electric dryer is to position it in a plenum located near the rear of the dryer cabinet.

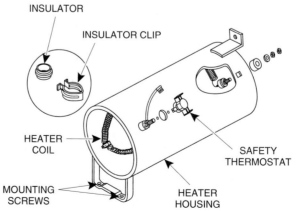

Figure 11-3. Another method of mounting a heating element is to position it in a steel tube located near the cabinet base.

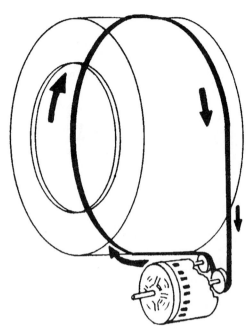

Figure 11-4. The belt of most dryers is routed around the drum, and tension is maintained by an idler pulley.

The two components most commonly found in the console of an electric dryer are the timer and the push-to-start switch. This allows the customer to set the dryer for the selected operating time and initiate the cycle with the push-to-start button. Figure 11-5 is an illustration of a typical dryer console. Some models, if they have more features, will contain more switches to allow more precise operation selection by the customer.

In some cases, the temperature selection may be accomplished by controls located on the console, but the most common method is the use of the *bimetal* or *thermo-disc thermostat* located somewhere near the exhaust path. The actual location and operating temperature of a bimetal thermostat depends on the manufacturer's design and also particular model year. It's not common to find a thermostat with a sensing bulb. The thermo-disc type of thermostat is also used as a safety thermostat, or *limit switch* as it is often referred to. Figures 11-6 and 11-7 give you an idea of one manufacturer's method of locating the bimetal operating thermostat and limit switches.

11.2 DRIVE MOTORS

The drive motor of an electric dryer operates on 120 volts. The 120/240-volt circuit that is supplied to the electric dryer is split, using one hot leg and the neutral to provide the 120 volts required to operate the drive/blower motor. Motors in clothes dryers are double-shaft

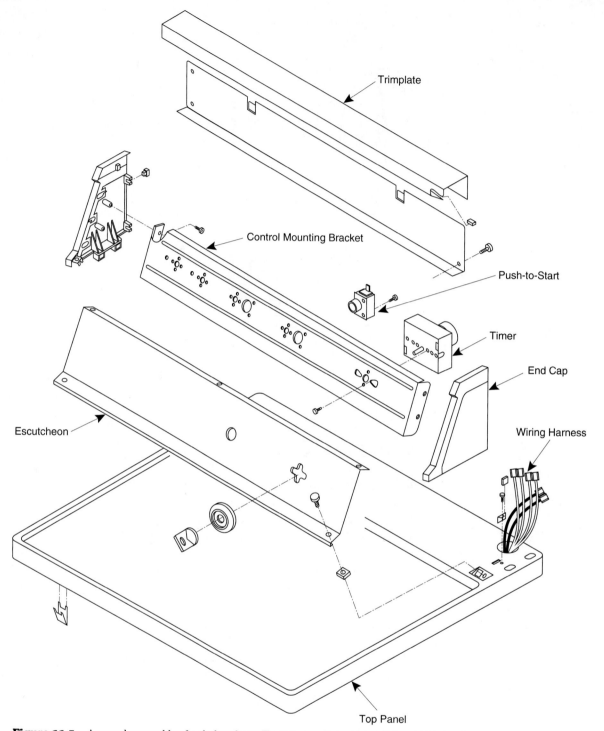

Trimplate

Control Mounting Bracket

Push-to-Start

Timer

End Cap

Escutcheon

Wiring Harness

Top Panel

Figure 11-5. A console assembly of a clothes dryer. The timer, push-to-start switch, and fabric selector switch is located in the console.

motors, allowing for a pulley that accepts the drum belt on one end and for a blower assembly attachment on the other end.

Figure 11-8 shows a typical clothes dryer motor assembly. As you can see, the motor sits in a cradle, and an idler assembly is used

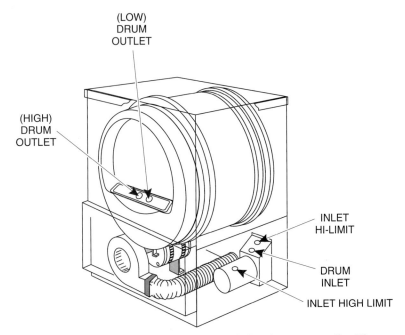

(LOW)
DRUM
OUTLET

(HIGH)
DRUM
OUTLET

INLET
HI-LIMIT

DRUM
INLET

INLET HIGH LIMIT

Figure 11-6. Operating thermostats used in clothes dryers are usually of the bimetal type and are located near the exhaust path.

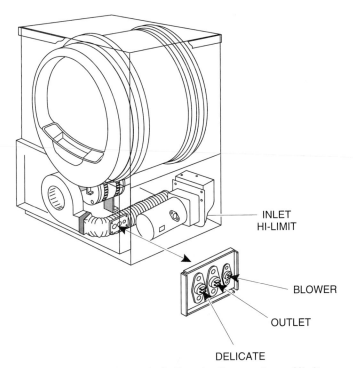

INLET
HI-LIMIT

BLOWER

OUTLET

DELICATE

Figure 11-7. Another method of locating the operating and limit thermostats in a clothes dryer.

to keep proper tension on the drum belt. In some cases the cradle may be bolted to the base of the dryer cabinet, and in some cases it may not be removeable from the base. It depends on the manufacturer of the equipment. Whether it is a removeable cradle, welded on, or

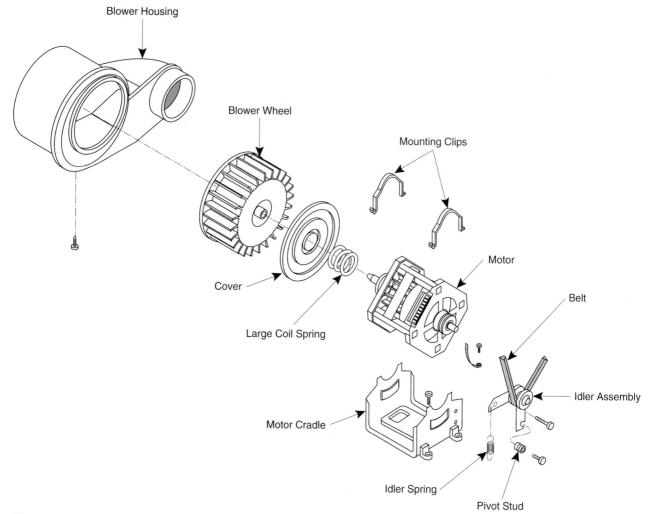

Figure 11-8. A drive motor in a clothes dryer uses two shafts to accommodate the belt pully and blower assembly.

part of the dryer base, a spring clip will be used to fasten the motor to the cradle. Figure 11-9 shows the mounting clips used to secure the motor.

In most cases both shafts of the motor will be threaded to accept the blower wheel and the motor pulley, and it's not uncommon to encounter a left-hand thread blower wheel. In some cases, the motor pulley and the blower wheel will be fastened to the motor shafts with set screws, while on some manufacturer's units you'll find the pulley fastened with a press fit. To remove a motor pulley installed in this manner, you'll need a special puller to get the job done right. One way around this problem would be to order a replacement pully with the new motor.

A dryer motor can be tested directly with the electrical system of the unit bypassed by using a simple test cord that you can build yourself. We'll be discussing a test cord of this type in Appendix B.

Dryer motors are single-speed motors that operate at a speed of 1,725 RPM, and a centrifugal switch is used to accomplish the start

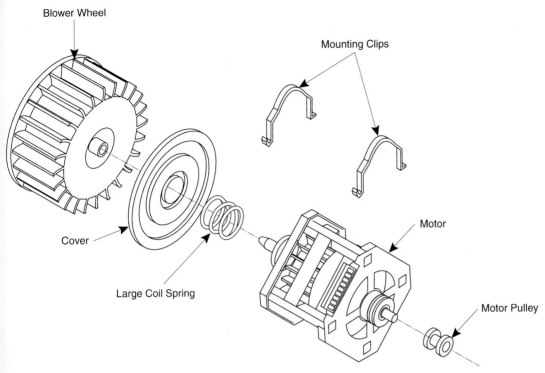

Figure 11-9.   The drive motor of a dryer is fastened to a cradle that is either fastened to or is part of the dryer base.

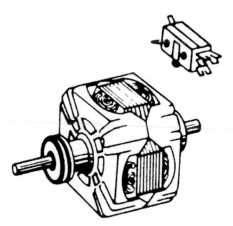

Figure 11-10.   A clothes dryer motor uses a centrifugal switch.

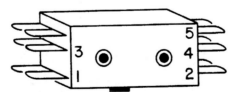

Figure 11-11.   The centrifugal switch on a clothes dryer can be replaced in the event of its failure.

and run. In most cases the centrifugal switch can be replaced without replacing the entire motor.

Figure 11-10 shows how a centrifugal switch attaches to a motor, and Figure 11-11 gives you a closer look at the switch itself. The same motor and switch can be used on a gas or electric dryer. The centrifugal switch, in many cases, is used not only to start and run the motor, but other terminals of the switch are used to provide a circuit to the heat source of the dryer. Wiring the unit in this way ensures that the heat will not be turned on unless the motor has reached its proper running speed and there is sufficient air flow through the air handling system of the dryer. Inside the centrifugal switch of a clothes dryer motor are separate sets of switching contacts. One set will make and break a circuit to the start winding of the motor, while the other set of contacts will make the circuit to the heat source once the motor has reached the proper operating speed.

11.3 GAS DRYERS

As far as cabinet construction, the gas dryer is almost identical in appearance to the electric dryer. Sometimes you may find that the front access panel on a gas dryer is slightly different due to the positioning of the burner assembly for service.

Figure 11-12 shows a simplified illustration of a gas clothes dryer. The burner is located underneath the drum in a steel tube. With this particular method of construction, the steel tube under the drum could also contain the heating element used in an electric

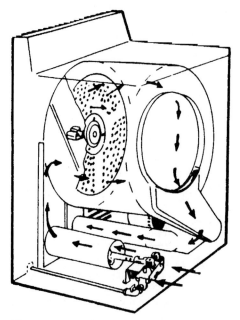

Figure 11-12. The air flow pattern of a gas dryer is similar to that of an electric dryer.

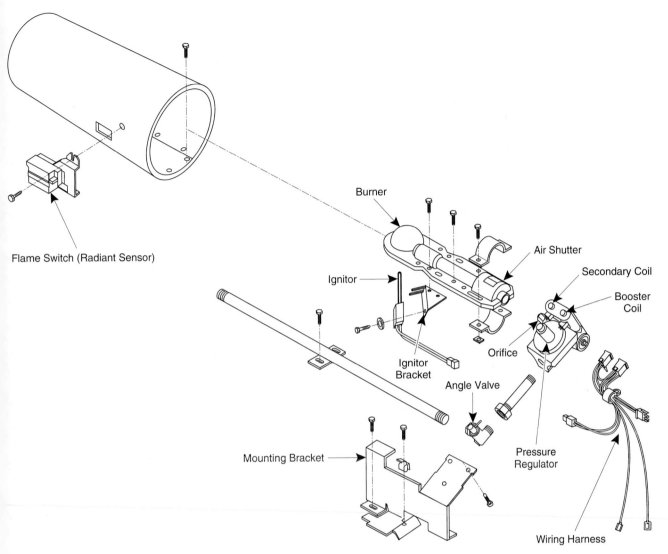

Figure 11-13. A gas burner assembly that uses a spark ignition device.

dryer. The air flow pattern is identical regardless of the heat source. As you can see, air is pulled into the dryer cabinet assembly at the front. Then it is drawn through the tube containing the burner before it is routed through the back of the dryer bulkhead, and forced over the clothes in the drum. After the warm, dry air has done its job of removing moisture from the clothes, the moist, warm air is then forced out the exhaust tube of the dryer assembly and out the vent system connected to the dryer.

Figure 11-13 gives you a closer look at the burner assembly as it fits into the steel tube. This assembly uses an electric spark device as an ignitor for the burner assembly. Older dryers may have a standing pilot and thermocouple system similar to that found on a water heater, but newer models use some type of electric ignition device.

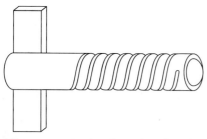

Figure 11-14. A carborundum glo-coil is also used to ignite the burner of a clothes dryer.

The *electric spark* method of igniting a dryer burner is one method used by manufacturers and the glo-coil method is another. A *glo-coil* is a resistive load that, when 120 volts is applied, glows bright red. The glo-coil is positioned in the path of the gas flow and when the gas valve allows gas to flow, ignition takes place. The glo-coil is a very fragile, brittle material known as carborundum. Figure 11-14 shows a carborundum glo-coil.

Whether the system uses a glo-coil or a spark device, the method of operation is the same. When the thermostat calls for heat, the ignition system is energized. When the operating temperature of the dryer is reached, the thermostat breaks the circuit to the burner assembly and the flame is turned off. The dryer continues to run and tumble the clothes until the temperature drops to the point where the thermostat again calls for heat. During a 30-minute time selection by the customer, the burner (or, in the case of an electric dryer, the heating element) is cycled off and on many times to maintain the desired operating temperature of the clothes dryer.

In addition to the proper venting of a gas dryer, another important factor the appliance technician has to consider is the proper operation of the burner. The air-to-gas mixture must be correct or the efficiency of the dryer will be affected. The proper adjustment of the burner is accomplished through the use of an air shutter that allows more or less oxygen into the burner assembly.

Figure 11-15 shows the shutter on a gas dryer burner assembly. The shutter is usually held in place with a locking screw, and a slotted section on the shutter allows for adjustment.

To ensure maximum efficiency, the technician must understand that the burner is operating at its peak when the blue flames of the burner are tipped yellow. Figure 11-16 shows three stages of shutter

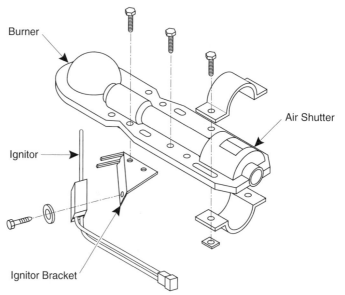

Figure 11-15. To ensure that a burner operates properly, an air shutter is provided for burner adjustment.

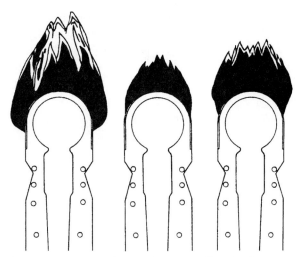

Figure 11-16. Insufficient air causes yellow tips that are too large, and not enough air causes a blue burner with no yellow tips.

adjustment on a gas burner. The illustration at the left is incorrect because it shows a flame with yellow tips that are too large, an indication that not enough air is being supplied through the shutter. The burner illustration in the center shows a burner that has the air shutter adjusted to a point where too much air is allowed, making the burner all blue with no yellow tips. The illustration at the right shows a correctly adjusted burner with small yellow tips at the end of the blue flame.

11.4 PROPER VENTING OF A CLOTHES DRYER

We'll be discussing the proper methods of installing a dryer in a later unit. But venting a dryer is so important and, if improperly done, causes so many problems with the operation of the dryer, it's worth a mention. If a vent is not proper installed or if it is clogged with lint, it restricts the flow of air through the dryer and as a result causes the limit switch to shut the unit down. When installing a dryer, you should avoid using the flexible plastic venting material that is so commonly

DIRECT EXHAUSTING THROUGH WALL

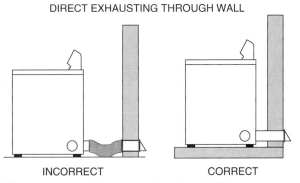

INCORRECT   CORRECT

Figure 11-17. A rigid vent, rather than flexible vent material, should be used when venting a dryer.

used. In some areas, this material has been outlawed. Whenever possible, use a rigid vent system of 4-inch pipe. Figure 11-17 shows the difference in using the rigid material and the flexible material.

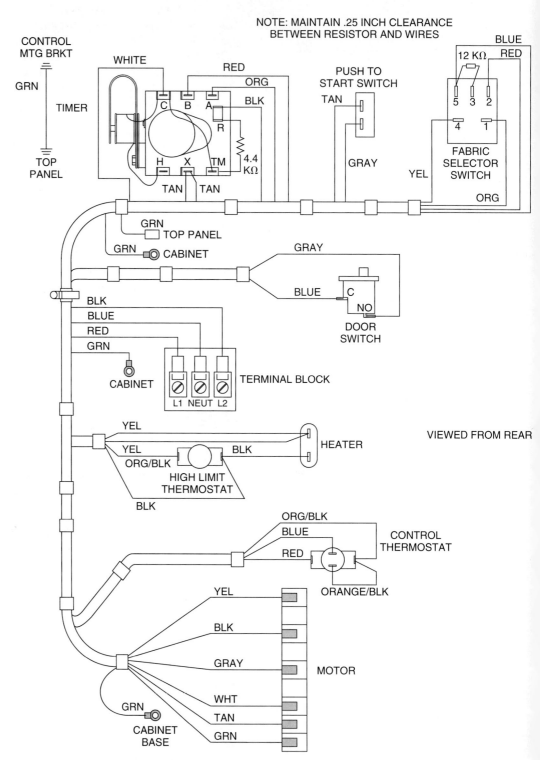

Figure 11-18. A pictorial diagram of an electric dryer.

11.5 ELECTRICAL SYSTEM

Like all appliances, the electrical system is responsible for nearly 80 percent of the problems the technician will encounter. When it comes to wiring diagrams, manufacturers use many different ways of illustrating the electrical system. Both the schematic and pictorial diagrams are used.

Figure 11-18 shows one method, used by White Consoidated Industries, to explain the routing of wires and the method of operation of the dryer. This diagram is for a Gibson electric dryer that operates on 120/240 VAC. The power is supplied to the unit at the terminal block located near the center of the drawing. The other components in the dryer's electrical system, such as the motor, timer, push-to-start switch, and others, are shown and the color coding of the wiring system is also illustrated.

CHAPTER ELEVEN   SUMMARY

Clothes dryers are relatively simple in their method of construction and operation. A clothes dryer may use electric heating elements, described as *ribbon-type* elements, or a gas burner. Modern gas dryers can be converted to operate on LP gas instead of natural gas. Regardless of the heat source, the mechanical operation of the dryer remains the same. The four systems used on a clothes dryer are the electrical system, the heat source, the air handling system, and the mechanical system.

Standard-sized electric dryers operate on 240 VAC due to the need for the high-amperage circuit of the resistance-type heating elements. A gas dryer operates on 120 VAC and plugs into a standard wall outlet. In the case of an electric dryer, its power cord is known as a *pigtail* and the 240-volt receptacle is much different in appearance. Drive motors in both gas and electric dryers operate on 120 volts.

Dryer motors operate at 1,725 RPM, are double-shaft to accept the belt pully on one end and the blower assembly on the other. It is common practice for the blower wheel to be threaded left-hand. A centrifugal switch is used to start and run the motor, and the switch can be removed and replaced without replacing the entire motor. The centrifugal switch breaks the circuit to the start winding after the motor reaches about 75 percent of its running speed, and in many cases the wiring to the heat source on the dryer is routed through the centrifugal switch. With wiring like this, the heat cannot be energized unless the motor is running at full speed.

Gas dryers are similar in appearance to an electric dryer. The gas burner may be ignited by a *spark device* or a device known as a *carborundum glo-coil*. In either case, the burner is cycled off and on during the run time selected by the customer.

Proper venting is very important. If a vent is improperly installed or if it becomes clogged with lint, it affects the operation of the dryer, causing it to shut off on the limit switch. Rigid venting rather than flexible venting should be used.

Manufacturers use both the schematic and pictorial diagram to show the electrical components of a system and the method of routing the wires.

CHAPTER TWELVE

Electric Ranges

LEARNING OBJECTIVES

After studying this Chapter, you will be able to:

1. Identify the electrical components of an electric range.
2. Identify the different styles of electric cooking equipment and differentiate between methods of construction.
3. Explain the method of operation of a self-cleaning oven.
4. Trace a simple electrical circuit in an electric range.

■ ■ ■

Beginning in the fall and throughout the winter, the appliance technician will find the number of calls on cooking equipment at higher levels than in the summer months. The electric range, wall oven, or cook top are among the types of cooking equipment that will require service. While the positioning of switches and thermostats and some differences in cabinet design may vary from one manufacturer to another, the basic operation remains the same: We use electrical energy and resistance elements to create heat. Controlling the operation of that heating element may be accomplished with variations in the control switching through the use of different types of relays and sensors. Or, in the case of surface units, they may be controlled by a voltage-sensitive switch in one case, by a current-sensitive switch in another. All electric ranges, regardless of design and manufacture, require a 240-volt circuit for operation of the resistance heating elements. A 120-volt circuit is rarely used for operation of a resistance-type heating element. One exception may be on some manufacturer's smooth top units when applying power to the small surface elements.

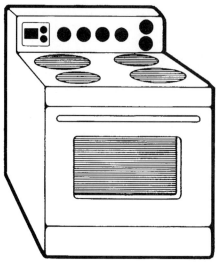

Figure 12-1. The free-standing 30-inch range is the most popular electric cooking appliance. Other sizes are available. Smaller units for apartments and larger units for the customer who wishes to have them.

12.1 THE FREE-STANDING ELECTRIC RANGE

The *free-standing range* is the most common style of electrical cooking equipment you'll encounter as a service technician. The most popular size is the 30-inch, model such as the one shown in Figure 12-1. This shows a standard oven that has four surface units and one oven cavity, in which both the baking and broiling processes are accomplished. With this type of unit, the bake element is located at the bottom of the oven cavity, and the broil element is located at the top.

Like an electric clothes dryer, the electric range requires a different receptacle, and a power cord known as a *pigtail* is used to supply power to the unit. The range pigtail is similar in appearance to the dryer power cord but must carry a higher level of energy since the dryer operates on a 30-amp circuit and the electric range requires a 50-amp circuit.

Figure 12-2 shows a typical electric range pigtail as it is connected to the main terminal block of the equipment. As you can see at the top of the drawing, a reading of 240 VAC will be shown by a voltmeter between points A and C. Correspondingly, points A and C at the bottom of the drawing show where the power cord would plug into the appropriate sections of the wall receptacle. You'll also notice that a reading of 120 VAC will be accomplished from the neutral leg to any of the two hot legs, identified as L1 and L2.

The higher voltage is required for operation of the surface units and bake and broil elements. The lower voltage is used to power convenience lights, clock motor, and, in some cases, the signal lights (sometimes referred to as *pilot lights*). The door locking motor or locking solenoid on a unit equipped with a self-cleaning system may

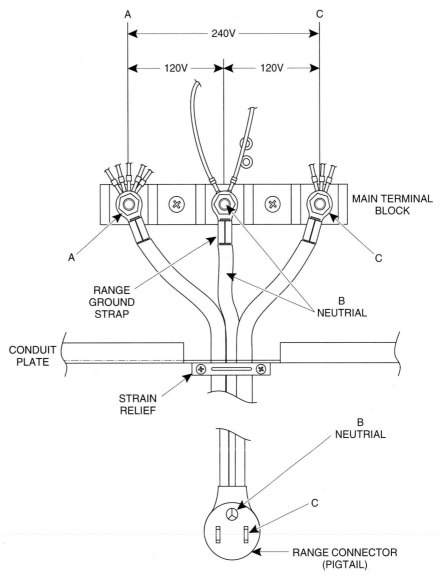

Figure 12-2.   A free-standing range uses a pigtail similar to that found on an electric clothes dryer. A 240-VAC circuit is required. *(Courtesy WCI).*

also be powered by the 120-volt circuit. This drawing also shows the terminal block used as a tie point for the wiring harness of the electric range. The terminal block on a free-standing unit is located on the rear of the range, and a small access panel must be removed in order to make the electrical tests described. The terminal block itself is made up of a plastic or bakelite body, in which threaded studs are fastened and nuts are used to assure a good electrical connection.

When it comes to mounting the switches and other related components that control the operation of an electric range, it is common to find them mounted in one of two ways: either near the front of the range, as shown in Figure 12-3, or in a backguard assembly near the back of the range, such as the one illustrated in Figure 12-4. In both cases, there are four surface unit switches, a selector switch that

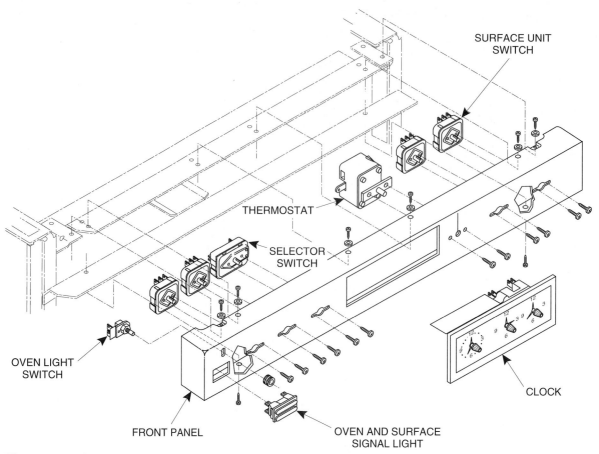

Figure 12-3.   On some ranges the controls are located near the front of the unit. *(Courtesy WCI)*.

allows the customer to select either bake, time bake, broil, or the self-cleaning mode if the unit is so equipped, a clock and timer assembly, oven thermostat, a switch that turns on a light inside the oven cavity and signal lights that indicate whether or not a particular component is energized.

You'll also note that Figure 12-4 offers a description of the method of operation of this manufacturer's equipment. In some cases, the customer has to select preheat as a mode on the oven selector switch. In that case, the bake setting would energize only the bake element. With this model, the manufacturer's explanation shows that, when the customer selects the bake mode on the selector switch, both the bake and broil element are energized until the temperature in the cavity nearly reaches the setting selected by the customer. At that point, the thermostat will energize only the bake element to maintain the desired baking temperature.

12.2   COOK TOPS AND WALL OVENS

In some homes you won't find a standard free-standing oven, but instead will see a *cook top* built into the countertop, and this type o

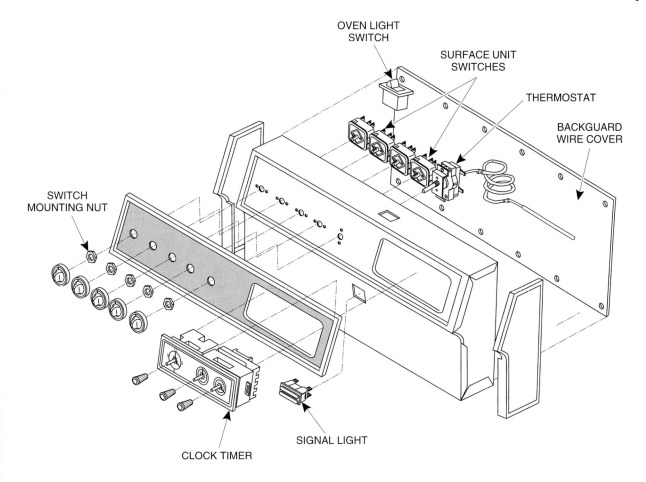

OVEN LIGHT SWITCH

SURFACE UNIT SWITCHES

THERMOSTAT

BACKGUARD WIRE COVER

SWITCH MOUNTING NUT

SIGNAL LIGHT

CLOCK TIMER

OVEN BAKE OPERATION

When the thermostat is turned to any bake temperature setting both the bake unit and the broil unit will heat up. This is called automatic pre-heat. The broil will only stay on until the oven temperature nears the dial setting, and will stay off for the remainder of baking time. The bake unit will cycle off and on to maintain temperature. The oven signal light will cycle on and off with the unit.

NOTE: The broil unit and the bake unit will come on together if the oven door is left opened for an extended period of time. Leaving the door open long enough for thermostat sensor bulb to cool down 100° F will turn both units on. (Thermostat goes back into pre-heat).

Figure 12-4.   Other models utilize a backguard assembly to mount the surface unit switches, clock and timer, lights, and oven thermostat. *(Courtesy WCI).*

unit is used in conjunction with a wall oven. Figure 12-5 shows one style of counter-mounted cook top. The switches and surface units used in a cook top assembly are identical to those in a free-standing combination unit.

Figure 12-6 shows a typical *wall oven*. Although it's referred to as a wall oven, you'll usually find it mounted in some kind of floor-to-ceiling cabinet assembly rather than fitted into an opening in a flat surface wall. The wall oven does not come equipped with any decorative side panels as the free-standing unit does. Only the front section of the cabinet has any trim or color to it.

Unlike the free-standing unit, the wall oven and the cook top are "hard wired" into the electrical system. A pigtail is not used. A flexible,

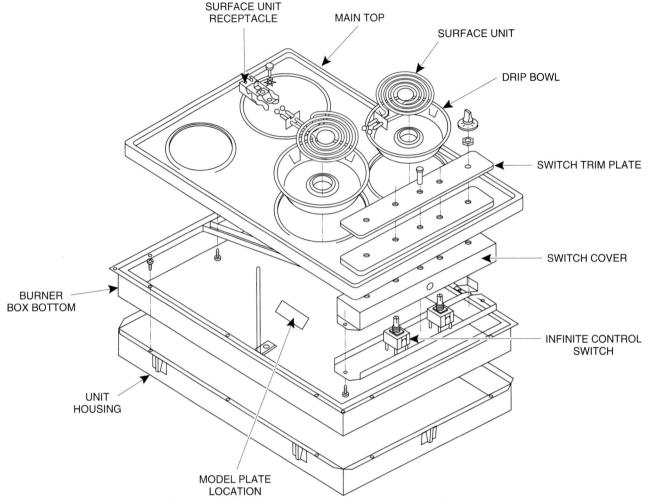

SURFACE UNIT
RECEPTACLE

MAIN TOP

SURFACE UNIT

DRIP BOWL

SWITCH TRIM PLATE

SWITCH COVER

BURNER
BOX BOTTOM

INFINITE CONTROL
SWITCH

UNIT
HOUSING

MODEL PLATE
LOCATION

Figure 12-5.   A cook top assembly, mounted in the kitchen countertop, is found in many residences.

armored cable is used as a wiring harness. A junction box, mounted either in the cavity behind the wall oven or in the cabinetwork under the countertop, will supply the electrical connections to these units.

Figure 12-7 shows one type of junction box you'll see when servicing wall ovens and cook tops. You'll note that, in this illustration, a four-wire system is used. In this case, L1 and L2 are still the two hot legs on the 240-volt circuit, and the white (neutral) leg is still used to supply a 120-volt circuit when used with one hot leg. The fourth wire in this system is the safety ground, a bare wire that is required according to some electrical codes. When a cooktop and a wall oven are used, each requires its own 120/240-volt 30-amp circuit.

Another style of electrical cooking equipment is known as the *drop-in range*. This unit is similar in construction to the free-standing unit but is does not sit on the floor, nor does it have a storage drawer as part of its cabinet. A *drop-in* has the surface units and the oven cavity in one unit, like the free-standing range, but this type of unit sits in a cabinet assembly such as that shown in Figure 12-8.

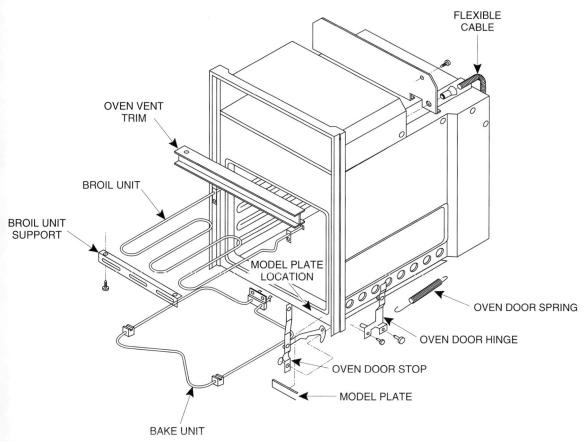

Figure 12-6. A wall oven is used in conjunction with a cook top. A unit of this type is hard wired to the electrical circuit.

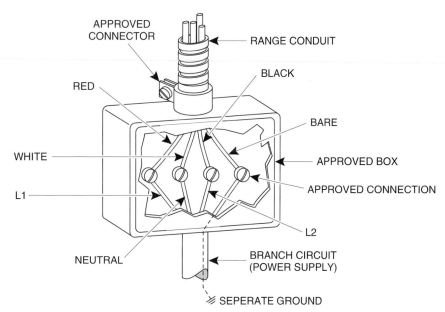

Figure 12-7. Some electrical circuits will have two hot wires, a neutral and a separate safety ground under some electrical codes. *(Courtesy WCI).*

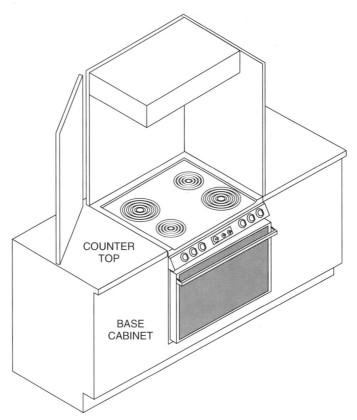

COUNTER
TOP

BASE
CABINET

Figure 12-8. A drop-in range is similar in construction to a free-standing unit.

12.3 OVEN THERMOSTATS

Many a thermostat in an electric range is similar in operation to a
refrigerator thermostat in that both types of controls utilize the capillary line/bellows method of making and breaking a switch inside the
control body. A thermostat of this type is shown mounted in the backguard section illustrated in Figure 12-4. The capillary line on an oven
thermostat will, in most cases, be longer than that of the refrigerator
thermostat because the sensing end of the line must be positioned
inside the oven cavity. In most cases, you'll find the end of the sensing
line held in place by metal clips, either near the rear or on one side of
the oven cavity. The purpose of the metal clips is to prevent the sensing
tube from damage and to enable it to maintain proper temperature.

Since many customers prefer to use the oven cavity as storage
for pots and pans, this positioning of the sensing tube can present problems. A technician sometimes finds that the temperature of the oven
isn't right because the sensing tube has been dislodged from its metal
clips and cannot correctly sense temperature. Keep in mind that the
sensing tube is designed to sense air temperature in a specific location within the oven cavity. If the sensing tube is either not in its proper position or is touching a metal surface, it can't perform its function.

When a customer selects a given temperature by turning the thermostat on, the thermostat is designed to maintain that selected temperature by making and breaking the circuit to the bake element. When calling for heat, the pressure in the tube is low, and the spring pressure inside the thermostat body keeps the contacts inside closed, allowing a complete circuit to the element. When the temperature increases to the thermostat set point, the spring pressure inside the switch body is overcome by the higher pressure inside the capillary line. The contacts inside the thermostat body are separated, breaking the circuit to the element. As the oven cools slightly, the pressure inside the tube decreases and the contacts again make contact, and the element is energized. A properly operating thermostat should maintain the selected temperature in the range of ±5° Fahrenheit if it's a newer, more modern unit or within a range of ±25° if it's an older unit. A selection of 350° would meant that the thermostat would break the circuit at about 355° and then make the circuit again at about 345°.

Not all temperature control systems are of the switch body/ capillary line type. You may find that in some instances a rheostat is used in conjunction with an oven sensor and a relay. This system does not use a spring pressure system to make and break the circuit to the element. Instead, the oven sensor is a thermistor that reacts to temperature. While there is no difference in appearance on the customer side of the backguard, removal of the range's back panel would reveal a control body similar to the type we've discussed, but instead of a capillary line, you'll see only a wiring harness connected to the switching assembly. This method of construction is commonly found on GE- and Hotpoint-built cooking equipment, which may be found under those brand names or under the brand name of a department store, such as JC Penney, that General Electric Company has agreed to supply.

A sensor of this type is shown in Figure 12-9. The probing end of the sensor can be seen in the oven cavity, and the bracket assembly attached to its body allows for mounting the oven cavity with two screws. To expose the wiring from the sensor and the relay used in this system, the back panel of the oven must be removed.

The temperature control system on an electric range usually operates in a range from 140° to 500° Fahrenheit. Most thermostats will also have a broil setting on them which will, when used without a selector switch and set to the broil setting by the customer, make a circuit to the broil element rather then the bake element. These types of thermostat will have a definite "click" feeling when the broil mode is selected. Other thermostats that don't have a definite switchover will still have a broil indication on their knob, but this type of control will be used in conjunction with a selector switch.

Calibrating a thermostat to a customer's satisfaction can be tricky. The customer may have an inexpensive and inaccurate thermometer, and the technician's task may be to convince the customer that the thermometer being used is not indicating correctly. If a

Figure 12-9. Instead of a capillary line thermostat, some units are equipped with an oven sensor that is used along with a rheostat and a relay to cycle the bake element on and off.

thermostat is off, it can usually be calibrated to make a break at the correct temperature. Always use an accurate high-temperature sensing device and allow the oven to cycle several times before adjusting. Some thermostats may be adjusted by removing the knob and locating the *adjusting screw*, commonly found in the center of the hollow shaft of the thermostat, while on others the "adjustment" consists of a slotted bracket on the back of the knob. With this type of adjusting system, the control itself is not calibrated, but the customer's setting changes to coincide with the operating temperature of the unit.

Frequently, when a customer reports that the thermostat on their oven is "off," the problem is often due to other factors related to improper customer usage, such as oversized pans that restrict air circulation in the oven cavity. Before you begin to adjust a thermostat, eliminate other posibilities that could cause the symptom of erratic or improper temperature.

12.4 SELECTOR SWITCHES

A *selector switch* on an electric range is used in conjunction with the oven thermostat. Some selector switches may simply allow the customer to switch between the bake and broil modes, while others may have several positions on them: bake, broil, time bake, or clean in the case of a unit designed with the self-cleaning feature. A selector switch of this type is shown in Figure 12-3.

12.5 BAKE AND BROIL ELEMENTS

The bake and broil elements used in an electric range are similar in construction to the defrost heaters discussed in previous units. A resistive wire is surrounded by a magnesium oxide (MgO) material, and there is an outer sheath made up of iron, nickel, and chrome. The magnesium oxide packed around the resistance wire actually performs two functions. First, it provides electrical insulation, and, second, it allows the conduction of heat to the outer sheath. Bake and broil elements will glow red-hot when voltage is applied and many different configurations and wattage ratings are used in different models. Figure 12-6 shows how the elements are positioned inside the oven cavity. Figure 12-10 gives you an idea of what you would see in regard to connections to an element when the access panel on the rear of the oven is removed.

When testing a bake or broil system, you can accomplish it one of two ways. Testing for voltage applied (240 VAC) to the elements or testing the elements for proper resistance. You can refer to Chapter Five for a review of the proper use of electric meters when testing electrical components. Figure 12-11 shows the use of an ohmmeter in testing a bake element for proper resistance.

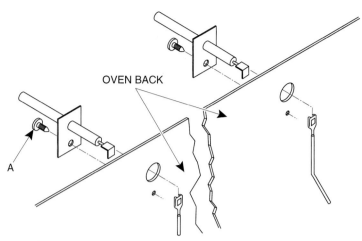

Figure 12-10. Bake and broil elements are mounted to the rear of the oven cavity, and the wiring is routed behind an access panel. *(Courtesy WCI).*

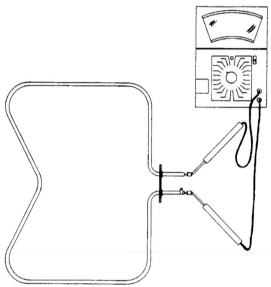

Figure 12-11. A bake element may be tested for proper resistance by using an ohmmeter.

Caution! *Never use an ohmmeter on a "hot" circuit. Doing so may cause damage to the meter.*

12.6 SURFACE UNITS

Surface units on a standard electric range are constructed in the same manner as a bake or broil element and are usually found in two sizes on any given model. It's common to find a 6-inch and an 8-inch diameter surface unit on a range, and in some cases, a manufacturer may use a 6- and a 10-inch diameter element. Some surface units may be mounted on hinged assemblies, to allow the customer the convenience of raising the unit and cleaning under them, and many units

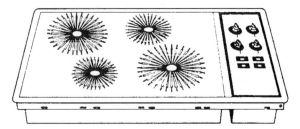

Figure 12-12. A smooth top cooking unit.

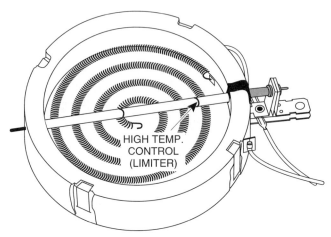

HIGH TEMP.
CONTROL
(LIMITER)

Figure 12-13. On smooth top units, a ribbon-type heating element is used.

can be found with plug-in elements. Figure 12-5 illustrates standard surface units.

In the case of a smooth top range or cook top unit, such as the one shown in Figure 12-12, a different type of surface unit is commonly used. Instead of the metal sheath construction method, a ribbon type element, similar to that used in an electric dryer and encased in an insulating material, is used. Figure 12-13 shows this type of heating element, which is mounted directly below the smooth top surface of the range.

Not all surface units require only two-wire connections. Some older ranges, such as GE/Hotpoint units, have dual element surface units. This type of surface element would require three-wire connections.

12.7 SURFACE UNIT SWITCHES

The purpose of the *surface unit switch* is to control the operating temperature of the surface unit according to the customer's varied needs. In some cases a high heat setting may be necessary, while in others a lower temperature is required. Achieving this control is accomplished in one of two ways: A switch is either to totally make and break a circuit to the surface unit, or to apply a small amount of voltage when low heat is required or a higher level of energy when high heat is required.

In the case of the make/break circuit, a switch that contains a bimetal heater is used. When the customer turns the knob and the switch completes the circuit to the surface unit, the bimetal device inside the switch is also energized. With voltage being applied to the bimetal inside the switch, it eventually heats up and warps, breaking the contacts that carry the power to the surface unit. When the bimetal cools, it returns to a position that again allows power to be supplied to the surface unit. When a customer selects a low setting, it doesn't take long for the bimetal to warp enough to cause the contacts to break. When a higher setting is selected, the bimetal travel distance is increased. As a result, voltage is applied to the surface unit for a longer period of time and a higher operating temperature is achieved. This type of switch is known as a *voltage-sensitive switch*, and the bimetal element inside it is wired in parallel with the surface unit, meaning that the bimetal is energized with an equal amount of voltage. Figure 12-14 shows a voltage-sensitive switch.

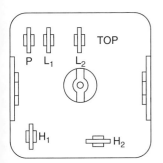

Figure 12-14. A voltage-sensitive surface unit switch must be mounted in a proper position.

Another type of surface unit switch that is used is known as a *current-sensitive switch*, also sometimes referred to as an *infinite switch*. This type of control uses a heating element inside the switch that is wired in series with the surface unit. All the load current flows through the heater before it gets to the surface unit. The result is that, on a low heat setting, voltage is supplied on an intermittent basis to achieve a low level of heat. On a higher heat setting, voltage is applied for a longer period of time. In other words, the resistance of the heating element inside the switch can be low or high, depending on the knob position selected by the customer. Figure 12-15 shows a current-sensitive switch.

Important Note: While a universal voltage switch can be used for a replacement, current infinite switches must be matched to the amp draw of the surface unit and are not interchangable from make to make.

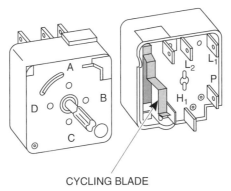

CYCLING BLADE

Figure 12-15. A current-sensitive surface unit switch operates differently than a voltage-sensitive switch but the wiring connections are the same.

With either style of switch, you must deal with five terminal connections: *L1* and *L2* allow for connection of 240 VAC into the switch assembly. *H1* and *H2* allow for wiring connections to the surface unit. L1 connects to H1 with one set of contacts inside the switch, and L2 connects to H2 with another set of contacts. The *P* terminal allows for a connection to a signal light (sometimes referred to as a *pilot light*). Only L1 travels out on this terminal, meaning that one side of the line is delivered to the signal light by way of a contact point inside the switch. In our example, the voltage delivered to our surface unit signal light is 240 volts. It's not uncommon to run into a unit in which the signal light is wired to one hot leg (through the surface unit switch) and to neutral, allowing for a 120-volt operation of the light.

It's common (but not always true all the time) that manufacturers break both sets of contacts inside the surface unit switch while it controls the operation of the surface unit. In some cases, only one set of contacts is broken during the cycling process. Breaking one hot leg still breaks the circuit and controls the operation of the surface unit, but it also means that one side of the unit is still hot, even if the unit is not fully energized. Always be careful when conducting voltage tests on an electric range. *Never assume that a surface unit switch breaks both sides of the line.*

Figure 12-16 shows a section of a wiring diagram related to a surface unit electrical circuit in which a current-sensitive switch is used, and Figure 12-17 illustrates a voltage-sensitive switch. In Figure 12-16, the heating element inside the switch is wired in series with the surface unit. In Figure 12-17, it is wired in parallel with the surface unit. Note the signal light circuit and its 240-VAC operation.

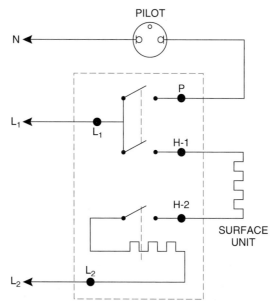

Figure 12-16. On a current-sensitive switch, the heating element is wired in series with the surface unit.

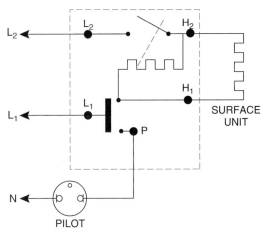

Figure 12-17. The bimetal heating unit in a voltage-sensitive switch is wired in parallel with the surface unit.

12.8 ELECTRICAL CIRCUITS

Many technicians are intimidated by an electric range when they remove the access panel. The reason for this is the vast array of wires that are necessary for the operation and control of the system's components. A careful look at the wiring diagram supplied with the appliance (or in the manufacturer's service manual) can help to eliminate problems and simplify your task. Refer to the diagram Figure 12-18.

First, note the legend on the diagram. It gives you valuable information. In this case, the legend lets us know the color coding of the various wires used in the circuits. The main colors are black, red, white, and green, and some wires may have a stripe either in green, blue, tan, or yellow. As an example, you'll note that the left rear surface unit has two wires connected that are designated as *RG*. This means that these wires are red with a green stripe. The stripe is sometimes referred to as a *tracer*.

The second step in effectively using any diagram is to be sure that you understand the component designations and how the switches work. For example, the four surface unit switches are designated as having connections at L1, L2, H1, and H2, as well as P. In the case of the oven thermostat, there is a designation for L1 (power into the switch), BRL (power out to the broil element), and BKE (power out to the bake element). The designation *PL* stands for pilot light and this circuit functions in the same manner as discussed on the surface unit switch signal light.

Once you have familiarized yourself with the diagram components, pick a circuit to trace. For our example we'll isolate the circuit to the oven lamp. First, trace it in the easiest way you can. In this case, that means starting with the center connection at the main terminal block and following the W (white) wire to the lamp.

Now, go on through the lamp and follow that wire all the way to the top of the diagram. You'll note that this wire is identified as B

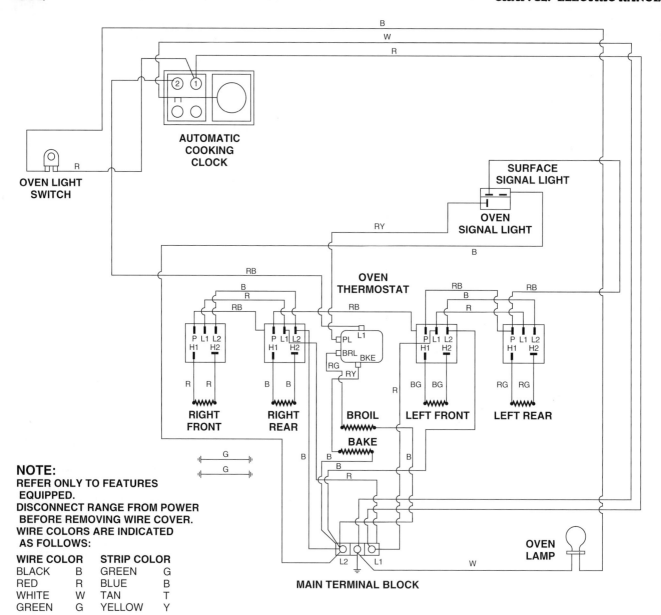

Figure 12-18. A simplified diagram of an electric range. *(Courtesy WCI).*

(black) and that it goes to one terminal of the oven light switch. Continue the circuit through the switch, exit on the R (red) wire, and you'll see that it connects to terminal #1 on the automatic cooking clock assembly. No work or switching is done here; the terminal is simply used in this circuit as a tie point, solely for convenience. Our color is still R (red) as the wire is shown going to the far right of the diagram, then down to a point where we can trace it to the left. It ultimately winds up connected to L1 on the main terminal block.

To test your understanding of this circuit, trace it from L1 on the red wire, to the tie point on the automatic cooking clock, to the switch, back to the lamp on the black wire, and finally back to the center post on the main terminal block via the white wire. Tracing this circuit from L1 through the switch, through the load (the oven lamp), and back to neutral will give you an understanding of the simplicity of this circuit: one side of the line through a switch, supplying power to a load, and completing the circuit back to neutral.

Once you've accomplished a simple 120-volt circuit, trace a 240-volt circuit. For our example, we'll use the circuit to the bake element of the range. As you can see, the bake element is identified as a resistive load, and a wire coded B is connected directly to one side of the bake element from L2. With the simplest starting point possible established, we'll trace the complete circuit.

Leaving L2 on the B (black) wire, trace it through the bake element, and follow the RY (red with yellow stripe) wire to terminal BKE on the oven thermostat. Since we know that our final objective is to supply 240 VAC to the bake element, it's logical that we leave the oven thermostat on terminal L1 so that we can ultimately return to L1 on the main terminal block, by whatever route the circuit takes us. Leaving L1 on the thermostat by way of the RB (red with black stripe) wire, we see that we are now connected to terminal #2 on the clock assembly.

As you can see, our only route to continue back to L1 is to leave terminal #1 on the clock assembly. A switching assembly inside the clock is not shown, but we know it must be there. We've already established on the previous circuit we traced that the R (red) wire leads back to L1 on the terminal block. The reason the wiring circuit is routed through the clock is to allow the customer the option of using the time bake feature.

Now test yourself as you did on the oven lamp circuit, and trace the circuit in the other direction. Start from L1 on the terminal block, go to terminal #1 on the clock, exit on terminal #2 on the clock, go to L1 on the thermostat, through the thermostat on BKE, then through the bake element to L2 on the main terminal block.

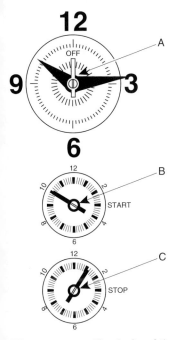

Figure 12-19. The clock and timer assembly on an electric range is used for timed baking and on a self-cleaning model, to set the cleaning mode. *(Courtesy WCI).*

12.9 SELF-CLEANING RANGES

Many electric ranges are equipped with a self-cleaning feature, meaning that the customer need not do anything beyond following the steps to initiate the clean cycle and then allow the range to do the work. While the exact procedures may vary slightly from one make and model of unit to another, some fundamentals are the same.

Self-cleaning is accomplished by energizing the bake element for a period of time (usually a minimum of two hours) and achieving a temperature of nearly 900° Fahrenheit, thereby turning any spills and heavy deposits in the oven cavity to ash. Further, much of the oven soil is decomposed since the intense heat removes the moisture from the material, then breaks down the hydrocarbons into smoke and gases. In the case of a heavily soiled oven, some smoke will be visible to the customer. In many cases, however, the customer won't

OVEN DOOR LOCK PARTS ILLUSTRATION

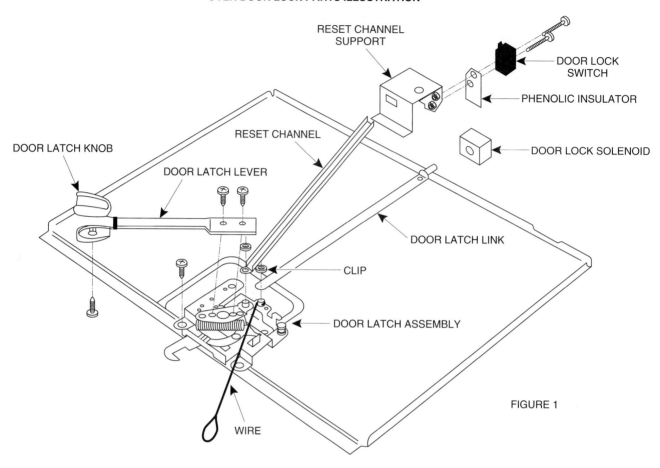

UNLOCKING LOCKED OVEN DOOR MANUALLY SELF CLEANING OVEN ONLY

If oven is cold, and door latch handle cannot be moved when momentary door latch switch is energized, the oven door may be opened manually by the following method:

  A. Use a strong straight wire similar to a straightened wire clothes hanger.

  B. Insert wire thru Vent Trim at door latch handle.

  C. Locate wire on latch mechanism thru Vent Trim using flash light (if necessary) and push latch toward
      rear of oven (direction of arrow) figure 1. With latch pushed back, move latch handle to unlocked position.

Figure 12-20.   During the clean cycle, the door locks to ensure the customer's safety. If a locking assembly fails, the technician must follow instructions to open the door manually.

see any smoke because this process allows the material to be vented out of the oven in a clear, odorless gas.

    The customer initiates the clean cycle by first selecting a start and stop time on a clock and timer assembly such as the one shown in Figure 12-19. In some cases, the owner's manual may direct them to raise a shield inside the oven door and move a lever to a locking position. Then the clean mode is selected on the selector switch and in some cases on the thermostat. Once the cleaning process has begun and a given temperature has been reached, the door remains locked until the cycle is complete and the oven has cooled down. Figure 12-20 shows a typical self-clean locking mechanism.

Once the clean cycle is complete, the customer can open the door and easily clean up any residue that may be left from the cleaning process. In many cases, the oven timer must be manually reset for a normal cooking mode after the clean mode has been selected and completed. Many customers forget to reset the timer and call for service with a "no bake" complaint. What the technician may find is that the unit need only be reset by pushing a timer knob (or pulling or turning it to a specific setting) to allow the range to be used in the baking mode.

CHAPTER TWELVE SUMMARY

In the winter months, appliance technicians find that the number of service calls for cooking equipment increases. Electric ranges operate on a 240-volt circuit because the higher level of energy is required to operate the resistance heating elements. In some cases, some small surface units on smooth top ranges may operate on a 120-volt circuit.

Electric ranges may be *free-standing units*, a *cook top* mounted in the counter may be used with a wall oven, or a *drop-in* range may be found in some residences. Free-standing ranges use a pigtail to connect to the 240-volt circuit, while cook tops, drop-ins, and wall ovens will usually be hard-wired. Some circuit may be four-wire circuits due to the electrical code.

Thermostats are frequently the capillary line/sensing bulb type. The sensing tube is positioned inside the oven cavity, and a long capillary line may be necessary due to the positioning of the thermostat body. Some thermostats may be calibrated if the temperature is not too far off the desired set point. A thermostat will usually maintain the desired temperature in the oven within ±5° Fahrenheit. Surface unit switches are either *voltage-sensitive* or *current-sensitive* control devices. One allows all voltage to be supplied to the surface unit, then makes and breaks the circuit. The other varies the amount of voltage delivered to the surface unit on a constant basis.

Bake and broil elements and surface units are constructed of a resistive wire surrounded by a magnesium oxide material, which is then surrounded by a sheath. Elements will glow red hot when fully energized. Many surface units used in smooth top ranges are constructed of a ribbon-type heating element embedded in an insulating material. The element is then positioned directly beneath the smooth top surface.

Self-cleaning ovens use extremely high temperatures to clean. During the cleaning cycle, a door locking assembly ensures the customer's safety by keeping the door locked until the unit is cool. In many cases, the clock and timer on a self-cleaning oven has to be reset manually before it can be used in the bake or broil mode.

CHAPTER THIRTEEN

Gas Ranges

LEARNING OBJECTIVES   **After studying this Chapter, you will be able to:**
1. Identify the different types of gas ranges used in residences.
2. Explain the difference between standing pilot and electric ignition systems on gas ranges.
3. Identify the fundamental components in a gas range.
4. Trace a simple gas range wiring diagram.

■　■　■

At a distance of 50 paces it's difficult to differentiate between a gas range and an electric range. The cabinet styles offered by gas range manufacturers are the same as those offered in electric ranges.

As an appliance technician, you can expect to find gas range repairs a relatively small part of your responsibilities. This is not due only to the fact that fewer gas ranges are built, but also because many gas ranges are simple in design and breakdowns don't occur as frequently. Older gas ranges are standing pilot systems, while newer, more modern units will be equipped with an electric ignition system for both the surface units and the oven burners. The most common combinations—although not the only ones—consists of surface unit burners ignited by a spark device and the oven burner ignited by a glo-bar device such as the one you will find on a gas dryer.

The actual method of getting the unit into operation varies from one manufacturer to another. Gas valves may be electrically operated in some units (operating voltage varies); in other cases, the gas valve may operate strictly on a pilot/heater pilot system and no electricity is required for main burner ignition.

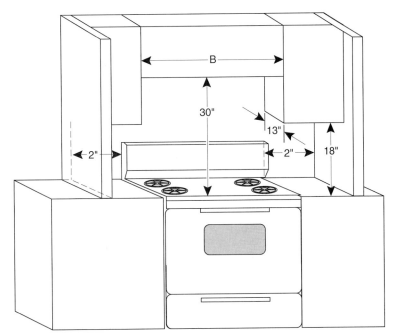

Figure 13-1.   The 30-inch gas range is the size found most often in homes using a free-standing unit.

13.1   FREE-STANDING OVENS

The free-standing gas range, like the electric range, is most popularly purchased in the 30-inch size, such as the unit shown in Figure 13-1. Gas ranges may be fit into the cabinet work without worry that they will do damage to the nearby wood materials due to the fact that they are insulated and properly designed for their application.

What serves as a storage drawer on an electric range usually houses the broiling tray on a gas range. The main oven gas burners will, in most cases, be mounted under the panel that makes up the bottom of the oven cavity. While in the bake mode, items placed on the oven racks are heated by warm air circulating around the oven cavity. In the broil mode, items mounted on the broiling tray are heated by a direct flame that is in close proximity. Some manufacturers don't follow this pattern. In those cases, the broiling burner is mounted to the roof of the oven cavity in the same fashion as the broiling element in an electric range. When a unit of this type (both gas and electric) is used for broiling, the manufacturer will usually specify that the oven door must be left partially open.

Figures 13-2 and 13-3 show the typical construction methods used in the gas range designed with two separate burners, one below the oven cavity bottom for baking and the other burner mounted in a position at the roof of the oven cavity for broiling.

In Figure 13-2, the bake burner is shown. The broil assembly is shown in Figure 13-3. You'll note that this particular bake unit uses a flame switch assembly. This means that this range utilizes a *standing*

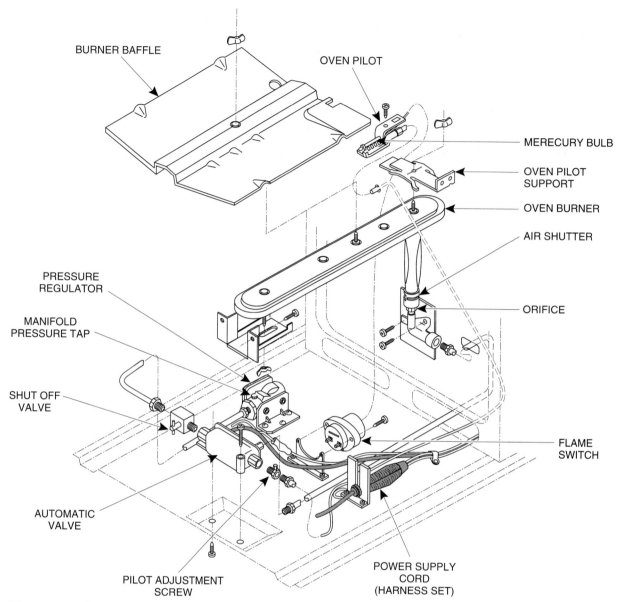

BURNER BAFFLE

OVEN PILOT

MERECURY BULB

OVEN PILOT SUPPORT

OVEN BURNER

AIR SHUTTER

ORIFICE

PRESSURE REGULATOR

MANIFOLD PRESSURE TAP

SHUT OFF VALVE

FLAME SWITCH

AUTOMATIC VALVE

PILOT ADJUSTMENT SCREW

POWER SUPPLY CORD (HARNESS SET)

Figure 13-2. The lower oven burner as it is found in a unit that uses a separate broiler unit.

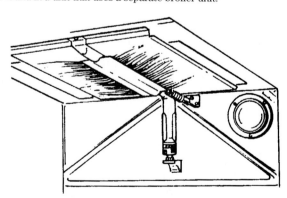

Figure 13-3. A broil burner located at the roof of the oven cavity.

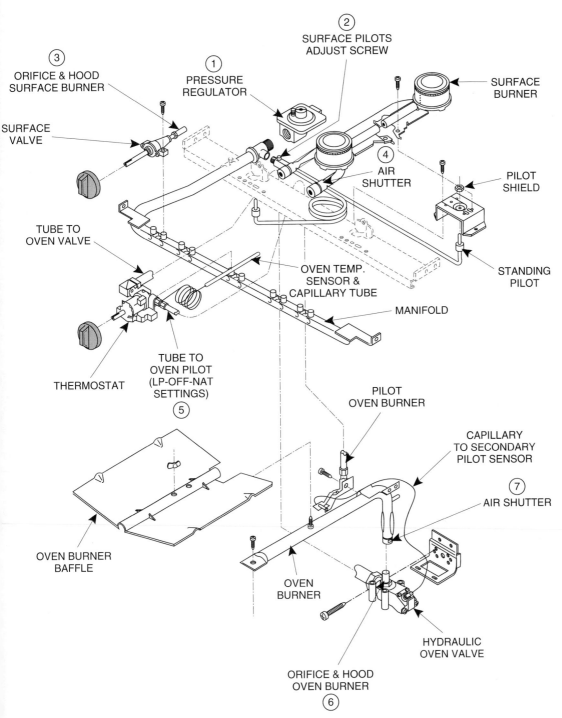

Figure 13-4. In some cases the oven burner works in both the bake and broil modes.

pilot, in which a sensing tube from the flame switch in the constant burning flame. When the customer turns the thermostat on, the electrically operated gas valve, which is wired in series with the flame switch, will be energized providing the pilot flame is established. If the flame has been extinguished, the flame switch will not allow the main gas valve to open and therefore prevents a dangerous situation in which large volumes of gas will be allowed to flow unignited into the oven cavity. All gas ranges (unless they are quite old) will use some method of safety in regard to the opening of the main gas valve. The flame switch method is only one of the methods used.

The broiler burner in Figure 13-3 shows the position of the burner assembly, along with a simulation of the gas flame that occurs when the burner is in operation. A burner of this type with totally exposed flame is relatively popular, although some manufacturers may use a burner assembly that does not show all the flame but instead emits only a bright red glow. This type of burner is similar in appearance to an infrared heating burner found in an industrial space heater.

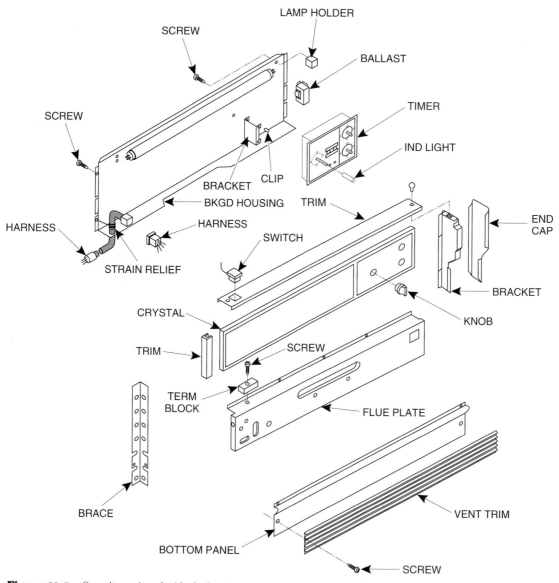

Figure 13-5.   On units equipped with clock and timer or convenience lamps, a backguard is used to support these components. *(Courtesy WCI).*

Figure 13-4 shows the configuration of the oven burner in a single-burner system. This illustration also shows the positioning of the main range manifold, which allows for surface valve installation. Other components such as the surface burners, oven thermostat, and pressure regulator are also shown.

Some free-standing units are equipped with clock and timer or a convenience lamp. In units such as these, a backguard assembly is used to support those components. The backguard assembly also serves as a support for the vent system of the range. A backguard is shown in Figure 13-5.

13.2 COOKTOPS AND WALL OVENS

Aside from the free-standing unit, a counter-mounted cook top assembly, used in conjunction with a wall oven, may be found in some homes. The wall oven requires no special venting because it is a gas rather than electric appliance, and the cook top unit may be mounted in the same fashion as the electric unit. The wall oven will appear exactly as the electric unit, while the cook top section may be slightly different in construction. The main difference being that the burners may be accessed simply by removing the cast iron utensil supports on top of the unit, then lifting the top to expose the burners and valves. A unit of this type is shown in Figure 13-6.

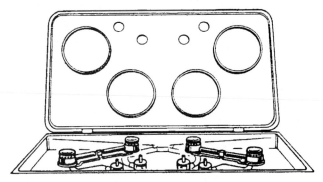

Figure 13-6. A gas cooktop assembly can often be opened for easy access to the burners.

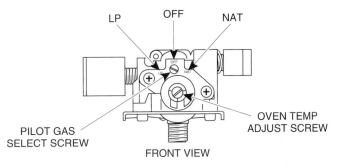

Figure 13-7. An oven thermostat that mounts directly to the range manifold. *(Courtesy WCI).*

13.3 OVEN THERMOSTATS

The oven thermostat on a gas range will most commonly be of the capillary sensing tube type. In some cases, the thermostat may actually control the flow of gas to the burner, or it may serve only to supply a heater pilot which, when satisfied the pilot flame has been established, will allow gas flow to the burner. A thermostat that actually controls gas flow either to the main burner or to a pilot assembly will be fastened to the manifold of the gas range. The body of the valve will allow for gas in and gas out connections.

A valve of this type is shown in Figure 13-7. As you can see, the body of this particular valve uses a threaded fitting to allow attachment to the manifold. This illustration also shows the common method of accommodating both the liquefied petroleum (LP) and natural gas installations. An adjusting screw ensures that the thermostat will properly operate the pilot on either system. On most modern gas ranges, no orifices must be changed to make the switch from natural to LP gas. The ranges are shipped set for natural gas.

Like electric range thermostats, the gas range thermostats can be adjusted to operate more precisely at the selected operating temperature through the use of an adjustment screw. The other method of adjusting the operating temperature of the range is to allow some adjustment in the knob. In this case, the knob is equipped with slotted openings that allow for repositioning of the knob indicating marks. Figure 13-8 shows an adjustable knob.

Some oven thermostats will not be directly connected to gas flow but will instead operate as a switch that either allows or doesn't allow for current flow to the electrically operated gas valve. In some cases, this type of thermostat will make or break a 120-volt circuit or

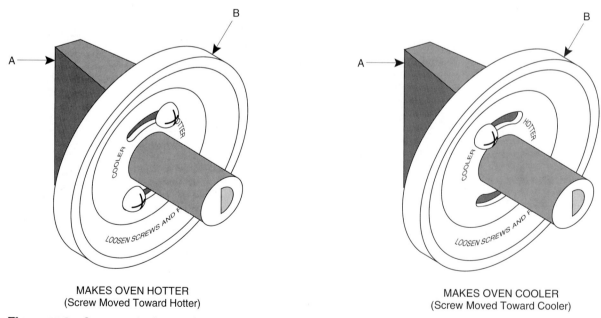

MAKES OVEN HOTTER
(Screw Moved Toward Hotter)

MAKES OVEN COOLER
(Screw Moved Toward Cooler)

Figure 13-8.   On some units the oven thermostat knob allows for temperature adjustment.

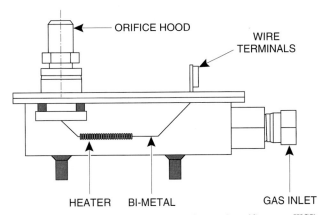

Figure 13-9. An electrically operated gas valve. *(Courtesy WCI).*

a step-down transformer may be used to deliver a lower amount of voltage to the gas valve. An oven thermostat of this type may also be used on glo-coil electric ignition systems. In these cases, the thermostat and the glo-coil are wired in a series circuit, which results in a lower voltage being delivered to the gas valve.

When an oven thermostat that acts as a switch is used, it supplies power to a valve such as the one shown in Figure 13-9. With this type of valve, the power delivered to it causes a warping action inside the valve, which in turn causes the opening of a seat inside the valve. This opening action allows the gas being delivered to the valve on a constant basis to flow through the valve body and on to the oven burner. The gas exits the valve assembly through the orifice hood, on which the burner assembly rests.

Other types of valves are used in conjunction with other types of thermostats. In the case of the thermostat that does nothing more than allow a standing pilot to be converted to a larger heater pilot, a valve such as the one shown in Figure 13-10 is used.

The tube that acts as a sensing element on this type of valve is positioned in a bracket that will allow contact by the heater pilot when the thermostat calls for heat. In a few seconds, the heat applied to the sensing element of the valve puts pressure on the operating

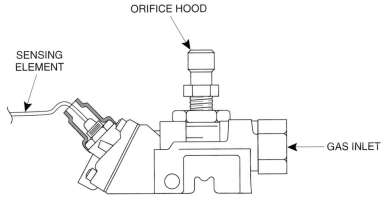

Figure 13-10. A gas safety valve that uses a sensing element positioned in a heater pilot.

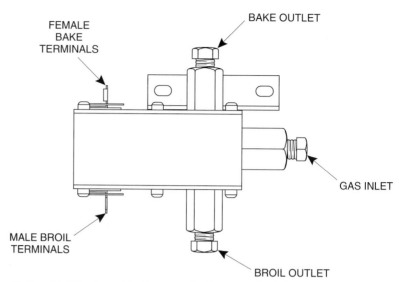

Figure 13-11. A combination gas valve that serves both the oven and broiler burner assemblies. *(Courtesy WCI).*

section of the valve. Once the pressure exerted by the now warmer sensing tube is sufficient to overcome the set spring pressure of the valve, the gas is allowed to flow through the valve body. The large supply of gas passes near the standing and heater pilot assembly, which is mounted near the main burner. As the gas begins to flow from the main burner, ignition is accomplished.

These aren't the only two styles of valves used on gas ranges. In some cases you find valves that serve separate broil and bake burners. As a result, these will have more than one outlet and also more than one place for wire connections. A combination valve is shown in Figure 13-11.

13.4 OVEN BURNERS

The oven burner on a gas range is a single tube assembly that allows the gas to flow around it in a U-shaped pattern. While the oven burner must be supported by some type of bracket, it simply sets on whatever type of valve supplies the gas. The burner must be adjusted for proper operation, and this is accomplished by varying the flow of gas from the supply valve and by adjusting for the proper amount of air. Without sufficient air for combustion, the flame will be yellow, carbonizing and not hot enough. With an excess of air, the flame will be too noisy and lift off the burner assembly. If the amount of gas being delivered to the burner is too high, the flame will go beyond the oven burner baffle and cause the food to burn. If the gas supply is too low, food will not be cooked properly and the oven temperature will be affected. The method for adjusting the flame is shown in Figure 13-12, and the air shutter adjustment is illustrated in Figure 13-13.

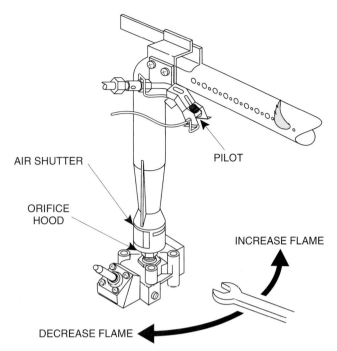

AIR SHUTTER

ORIFICE
HOOD

PILOT

INCREASE FLAME

DECREASE FLAME

Figure 13-12. The oven burner is adjusted for proper operation by adjusting the orifice hood on the safety valve.

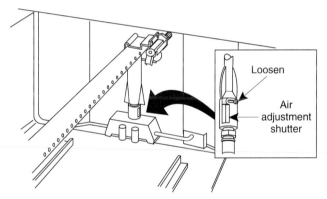

Loosen

Air
adjustment
shutter

Figure 13-13. An oven burner must also be adjusted for the correct amount of air so as to ensure efficient operation of the unit.

These same two adjustments are made when a gas range is converted to LP gas operation from natural gas operation. The orifice hood is screwed down closer to a shutoff position because LP gas is delivered at a higher pressure than natural gas. This higher pressure would result in a flame that is much too high if the valve orifice hood is not adjusted. In the case of the air shutter, it is opened as wide as possible for LP gas operation. Figure 13-14 shows the slight yet important difference in the adjustment of a valve orifice hood when a gas range is used on an LP gas system rather than with natural gas.

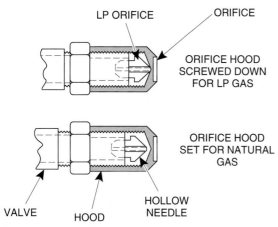

LP ORIFICE   ORIFICE

ORIFICE HOOD
SCREWED DOWN
FOR LP GAS

ORIFICE HOOD
SET FOR NATURAL
GAS

VALVE   HOOD   HOLLOW
NEEDLE

Figure 13-14.   The orifice hood is also used to ensure
proper operation of the range on LP gas.

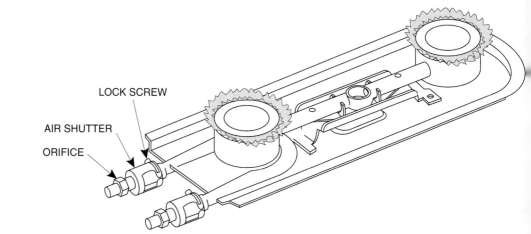

LOCK SCREW

AIR SHUTTER

ORIFICE

Figure 13-15.   Surface unit burners are adjusted for proper air mix through the use of an air shutter.

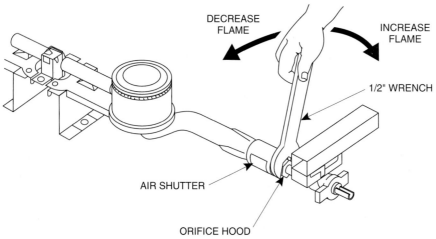

DECREASE
FLAME

INCREASE
FLAME

1/2" WRENCH

AIR SHUTTER

ORIFICE HOOD

Figure 13-16.   A surface burner may also be adjusted for proper operation and for LP operation
through the use of an orifice hood adjustment.

13.5 SURFACE BURNERS

Surface burners are supported by brackets on the oven top and slip over a valve hood assembly in the same way as the oven burner. A single pilot assembly, which may be either electric or standing, is positioned half way between two burners (one front, one rear) and a flash tube accomplishes the ignition of the gas that is allowed to flow to the burner head when the gas valve is turned to the on position. Like the oven burner, the surface burner must be adjusted for both gas volume and proper oxygen supply. One type of burner is shown for the purpose of illustrating air adjustment in Figure 13-15, and another style of burner is shown in Figure 13-16 illustrating proper gas volume supply.

13.6 GLO-COIL IGNITION SYSTEMS

While older gas ranges use standing pilot systems, newer units will use some method of igniting the burners with an electric ignition system. Two types of electric ignition systems, the glo-coil system and the spark device, may both be found on a gas range.

The glo-coil method is used on oven burner ignition. The *glo-coil igniter*, also referred to as a *glo-bar*, is the same type of igniter used in gas dryers. It is manufactured of a resistive material that will heat to an extremely high temperature when power is applied. This type of igniter glows bright red when it is energized. Styles vary somewhat, as is shown in Figure 13-17.

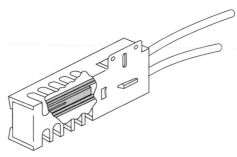

NORTON GLOWBAR

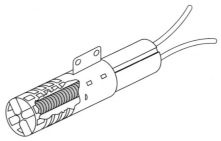

CARBORUNDUM GLOWBAR

Figure 13-17. A glowbar igniter is used on modern gas ranges. *(Courtesy WCI).*

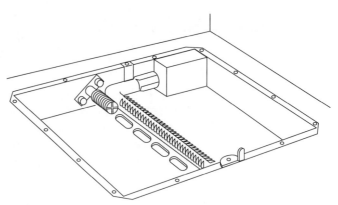

Figure 13-18. A glo-coil igniter is located near the gas flow to the oven burner. *(Courtesy WCI).*

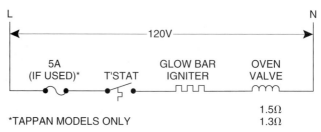

*TAPPAN MODELS ONLY

Figure 13-19. A simplified wiring diagram for a gas range with an electric ignition system.

The operation of the glo-bar ignition system is relatively simple. The igniter itself is located in close to the burner. (See Figure 13-18.) Figure 13-19 shows a simplified electrical diagram of the glo-bar ignition system.

13.7 SPARK IGNITION SYSTEMS

A *spark ignition system* consists of a lighting module that contains a capacitor. When power is applied to the unit, it discharges a high voltage. The discharge wire from the lighting module is positioned near metal. When the high voltage from the capacitor causes current flow down the wire, an arc occurs because the high voltage will cause electricity to jump from the wire to ground, the metal bracket that supports the electrode wire. Once the brief discharge of high voltage creates the arc, the circuit breaks and the discharge capacitor once again delivers the high-voltage spark. Most intermittent spark devices are designed to create the arc two or three times per second and are commonly found on surface burners. In some cases, a spark ignition device may be used to ignite an oven burner. In this instance, the spark ignition system lights a pilot, which in turn causes the main gas valve to open and deliver gas to the main burner. Figure 13-20 shows a typical spark module found in gas ranges.

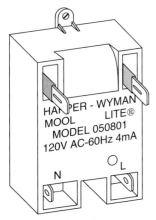

Figure 13-20. A spark ignition system on a gas range uses an igniter module that contains a capacitor.

CHAPTER THIRTEEN   SUMMARY

Gas ranges, similar in appearance to electric ranges, are available as free-standing units and as wall ovens used in conjunction with a cook top fitted into the countertop. Since gas ranges are properly designed and constructed, no special venting considerations are necessary.

Many gas ranges will use only one main burner both for baking and for broiling. In this type of unit the single burner is located under the oven cavity, and a baffle is used to allow the burner to burn and the heat to migrate to the baking area. Some units use a separate burner for broiling. In this case the oven cavity is used for broiling, and the broiler burner is mounted on the roof of the oven cavity.

Thermostats are often the capillary sensing bulb type. Some are directly mounted to the gas manifold of the range while some are a switch. A thermostat is always used with a safety valve to ensure that a large amount of gas doesn't flow into the oven without ignition occurring. Safety valves may be operated electrically, or they may be activated through the use of a sensing element positioned in a heater pilot assembly. In the case of a separate broil and bake burner, a combination valve may be used.

Most oven burners are single burners that allow the gas to burn in a U shape and must be adjusted properly for both the correct amount of air and the correct volume of gas. The adjustment of the orifice hood and the adjustment of the air shutter of an oven burner must also be done when the unit is being converted in its setup from natural gas to LP gas operation. All units are shipped from the factory set for natural gas operation.

Surface unit burners use a flash tube to accomplish the ignition process. The flash tube runs from the pilot source, either standing or spark device, and ignition takes place when the gas is allowed to flow to the burner. One pilot serves two burners, one front and one back on one side of the range. A second pilot serves the front and rear burners on the other side of the range.

The *glo-bar*, or *glowcoil*, ignition device is commonly used on newer gas ranges. The glowcoil is a resistive material wired in series with the gas valve. As the glo-coil heats up, its resistance drops, allowing a small amount of voltage to get to the gas valve. The gas valve won't be energized until the glo-bar has been energized long enough and has become hot enough to ignite the gas flow to the burner. *Spark ignition* devices are commonly found on surface unit burners but may also be used to ignite a heater pilot on an oven burner assembly. The spark igniter is actually a capacitor that is wired to discharge to ground, creating a spark near the gas flow to the burner.

Microwave Ovens

After studying this Chapter, you will be able to:
1. Explain how food is cooked by microwaves.
2. Identify the basic components in a microwave oven.

■   ■   ■

Since the development of the microwave oven, one of the major responsibilities of the appliance technician has been to explain the operation of the appliance to the customer to convince them that the unit is not a threat to their personal safety. The reason this situation exists is due mainly to misinformation and false assumptions that microwave signals generated by a microwave oven compares to those signals used in military and powerful radar equipment.

In addition, signs posted in restaurants warning pace maker wearers that a microwave oven is in use have contributed to the customer's unwarranted concerns about the safety of the appliance. In theory, the signal generated by a microwave oven could disrupt the electrical impulses that control a pacemaker, but the wearer would have to be extremely close to the unit and there would have to be a defect in the door seal.

Couple this information with the facts that pacemakers use a frequency far from the the microwave oven signal and that pacemakers are also shielded to prevent any outside interference from outside sources, and you can shatter one of the myths surrounding the mysterious microwave oven.

14.1 MICROWAVE OVEN THEORY

Microwaves are electromagnetic waves of radiant energy. To understand how they accomplish the cooking process, you first have to accept the fact that they exist. One common comparison to microwave energy used in domestic ovens is the broadcast system used in TV and radio systems. There are some similarities. Both systems use electrical energy to send out a signal in a radiant pattern from its source, and both systems generate this signal on a given frequency.

The difference in the two systems is that the broadcast signal is received by a device, such as your radio or TV set, and it converts the signal into sound or a picture. The signal in a microwave oven is used to create friction, which in turn creates heat. The other thing to consider, of course, is that the power level of the TV or radio broadcast station is many thousands of watts, while the power level of a microwave oven is only somewhere in the neighborhood of 400 to 1,200 watts depending on the model and the manufacturer.

The term *hertz* is important (discussed in the chapter on electrical fundamentals). *Hertz* means cycles per second. When we talk about broadcast systems and microwave ovens, we modify the term: *kilohertz*, meaning thousands of cycles per second, and *megahertz*, meaning millions of cycles per second. An AM radio station may generate a signal that operates at 50,000 kilohertz, 50,000 cycles per second. A TV signal may be emitted at a frequency of 204 megahertz, or in simple terms, 204 million cycles per second. This concept can be a difficult one to comprehend—to think that equipment can generate a

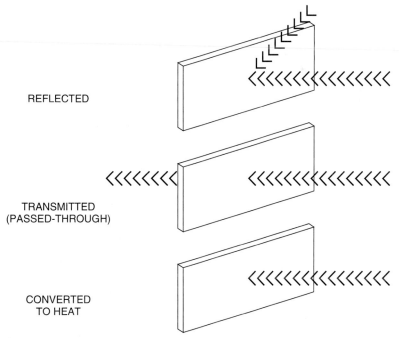

REFLECTED

TRANSMITTED
(PASSED-THROUGH)

CONVERTED
TO HEAT

Figure 14-1. A microwave can be reflected by a shiny surface, passed through glass, plastic, or paper and absorbed by other materials such as food and water.

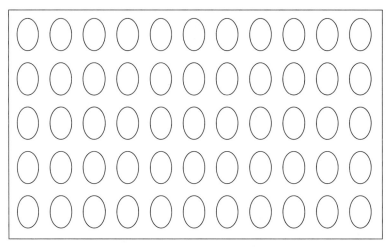

Figure 14-2.   Molecules at rest in a small section of a food product.

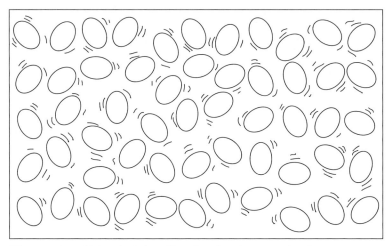

Figure 14-3.   Molecules being affected by the microwave signal.

signal that can change from a positive peak to a negative peak 204 million times per second. But it's something you have to accept to understand how a microwave oven accomplishes the cooking process. A microwave oven signal is generated at 2,450 megahertz— 2,450 million cycles per second.

Another comparison to make to better understand how a microwave oven cooks food is to consider the light wave. A light bulb emits light waves, and there are three things that can happen to a light wave. First, they can be reflected. If you hold a shiny piece of metal near a light bulb, you can see this happen. Second, it can be transmitted, or passed through, certain materials such as glass or plastic. Third, it can be absorbed. If you cover a light bulb with one paper towel, some of the light will be absorbed. If you add more layers of paper towels, you'll eventually absorb all the light waves emitted by the bulb. A microwave shares these characteristics. It can be reflected, transmitted, or absorbed, as shown in Figure 14-1.

Next consider that all food is made up of molecules. Since all molecules are made up of atoms, which are electrical in nature, the molecule fundamentally has a "north" and a "south" or a negative and a positive. If nothing is disturbing this electrical alignment, as is the case with the molecules in Figure 14-2, then they are at rest. And the molecules are at rest with their "north" pointing north and their "south" pointing south.

The microwave cooking process takes place when the signal generated by the unit interrupts this rest and causes the molecules to vibrate in search of their proper pole. As the molecules try to align themselves electrically, and the signal pulsates back and forth from negative to positive, the molecules rub against each other. As they rub against each other, they create friction, which creates heat. This attempt at alignment is happening at a rapid pace—2,450 million times per second. Figure 14-3 illustrates molecules being affected by a microwave signal.

Understanding these facts can enable you to shatter two of the most widely held myths about microwave ovens. First, that they cook from the inside out, and second, that any metal in a microwave oven causes problems. Figure 14-4 can help you explain the reasons these myths are not true. It shows a microwave oven cavity and the microwaves being generated by what is known as the *magnetron tube*. The microwaves are directed along a wave guide, and a component called a *stirrer* (sometimes referred to as an *antenna*) directs the microwaves in a pattern around the cavity so that the cooking process is accomplished as evenly as possible.

As far as metal goes, it's true that certain things in a microwave oven, such as crumpled aluminum foil or porous metals, will allow an arc to occur and cause problems in the operation of the unit. And smooth metal surfaces, such as stainless steel or aluminum, serve to cause the microwave to be reflected. Almost all microwave manufacturers use a metal cavity. Some may be painted (us-

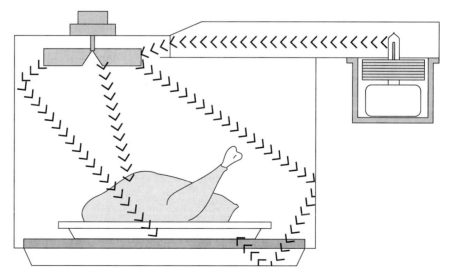

Figure 14-4. Microwaves penetrate up to a depth of about 2 inches. The inner layers of the food cook through the processes of conduction and convection.

ually a white or almond color) while others, such as Amana, may use a stainless steel cavity that appears as a shiny, polished metal surface.

In regard to the inside-out misconception, microwaves penetrate up to a depth of about 2 inches. After that, basic thermodynamics (the study of heat transfer) takes over to cook the inside of the roast or chicken. Specifically, heat always moves from a warmer surface to a cooler surface, and two of the ways it moves are by conduction (movement through a solid) or convection (movement through moisture). Since food is made up of solids and liquids, the heat created by the friction on the outer layer of the food travels to the center and the entire item is thereby cooked.

Also remember that we could enclose the turkey completely in glass or plastic, and still have it entirely cooked. Microwaves are transmitted (passed through) glass, plastic, and paper. The absorption of the microwave energy takes place when the molecules of the food react to the electromagnetic waves.

Another reason your customer may have doubts about the safety of the microwave oven is due to another term associated with the cooking process. The term *nuke*, relating to the nucleus of the atom, causes some people to be concerned because they don't understand electrical fundamentals and the idea of negative and positive in regard to molecules.

Microwaves are also thought by some to be an accumulating

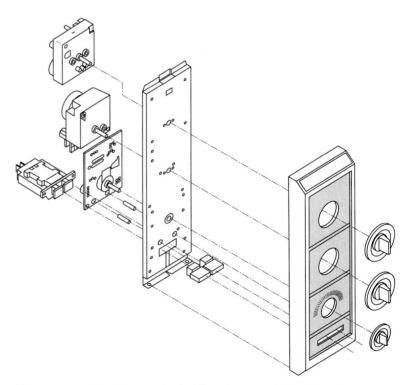

Figure 14-5. The electromechanical timer is commonly found on microwave ovens. *(Courtesy WCI).*

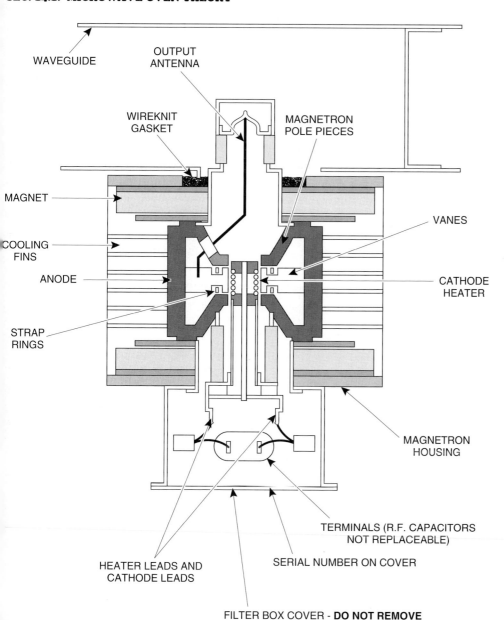

WAVEGUIDE

OUTPUT
ANTENNA

WIREKNIT
GASKET

MAGNETRON
POLE PIECES

MAGNET

VANES

COOLING
FINS

ANODE

CATHODE
HEATER

STRAP
RINGS

MAGNETRON
HOUSING

TERMINALS (R.F. CAPACITORS
NOT REPLACEABLE)

HEATER LEADS AND
CATHODE LEADS

SERIAL NUMBER ON COVER

FILTER BOX COVER - **DO NOT REMOVE**

Figure 14-6. A cutaway illustration of a magnetron tube.

type of energy, similar to X-rays. The belief is that the rays are being "absorbed" by the food, thereby adding something to the product. Both concepts, as you can now understand, are false.

14.2 MICROWAVE OVEN COMPONENTS

Like all appliances, the microwave oven uses a timer to control the other components and to perform the work selected by the customer. A timer on a microwave oven can range from a simple mechanical timer that times down from a set time to 0, to a printed circuit board system that is touch-sensitive and allows the programming of several features. Somewhere between the simplest and most complicated is the electromechanical timer, one similar to those we've already discussed that are switching devices driven by the own motors. A timer of this type, as well as a manufacturer's method of mounting it to allow the customer to set it, is shown in Figure 14-5.

The *magnetron tube*, or *mag tube*, as it is often referred to by technicians, is known as the heart of the microwave oven. It's the device that actually generates the microwave energy when high DC voltage (4,000 volts) and a low AC voltage (approximately 3 volts) is applied to its terminals. The magnetron is best described as a vacuum tube. The main body of the unit consists of an *anode*, a *cathode*

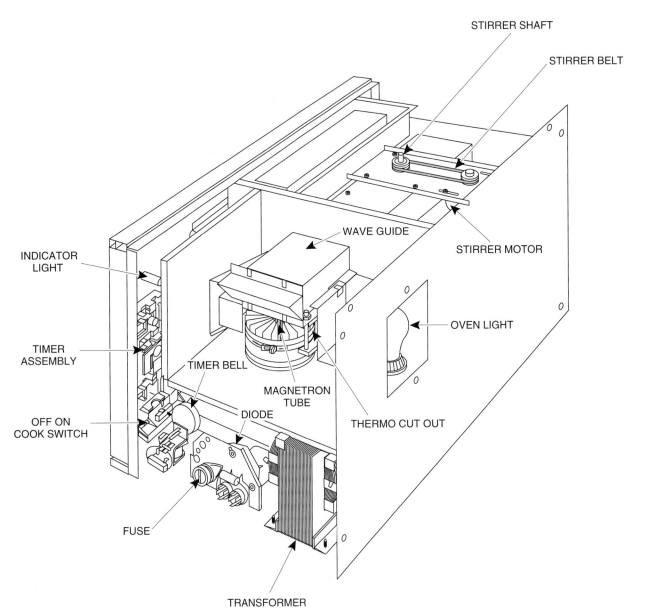

Figure 14-7. One manufacturer's design of microwave oven and the location of the components.

(also referred to as the *filament*), cooling fins, permanent magnets mounted on either side, and an antenna to radiate the microwave power the unit generates. Figure 14-6 shows a cutaway view of a magnetron tube.

The high and low voltage necessary to operate the magnetron tube so that it can generate RF (radio frequency) power comes from the power transformer. Some manufacturers use two separate transformers, one step-up and the other step-down, to supply the power, but in many cases you'll see one transformer with separate sets of secondary windings used to accomplish this. The primary of the transformer accepts 120 volts AC, and the secondary windings will allow an output of 2,000 volts AC and approximately 3 volts AC. A transformer, along with other basic microwave oven components, is shown in Figure 14-7.

As mentioned previously, the magnetron tube operates on 4,000 volts DC. While the transformer puts out 2,000 volts AC, before this energy is applied to the mag tube, it is routed through a capacitor and a diode (a rectifier), which make up what is known as a *voltage doubler circuit*. The voltage doubler circuit accepts the 2,000 volts AC, doubles it to 4,000 volts, and converts it from AC to DC.

Another important component in a microwave oven is the *cooling fan*. Its main function is to cool the magnetron tube, but it may also be used to drive the stirrer if that component is not driven by its

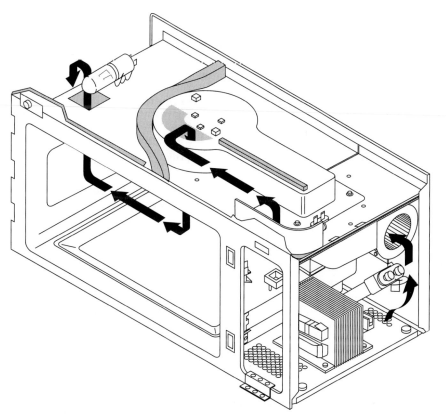

Figure 14-8.   The air flow path of the cooling fan on a microwave oven.

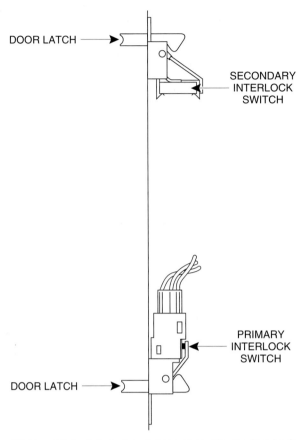

Figure 14-9. Interlock switches activated by the door latch assemblies prevent the oven from operating if the door is not closed properly.

own motor or motor and belt system. In some cases you'll find the cooling fan using a propeller blade, while in others it may be a squirrel cage blade. One method of magnetron cooling is shown in Figure 14-8. Although there is air flow through the magnetron tube body and out of the oven, there is no danger of microwave leakage due to the cooling fan system.

While the cooling fan ensures the safe operation of the unit by keeping the magnetron cool, the *door switches* that prevent the unit from operating unless the door is properly closed and sealed are also safety components. The three switches commonly found in a microwave oven are (1) the primary interlock switch, (2) the secondary interlock switch, and (3) the interlock monitor switch. In the event of a failure of any of the switches, the microwave oven will not operate until repairs are completed. The operation of the switches are directly related to proper operation of the door, as shown in Figure 14-9.

The *stirrer assembly* in a microwave oven serves to promote even cooking. All microwaves will have *hot spots*—areas in which cooking occurs faster than in others—and the stirrer eliminates some of these conditions by directing the microwaves in a more even pat-

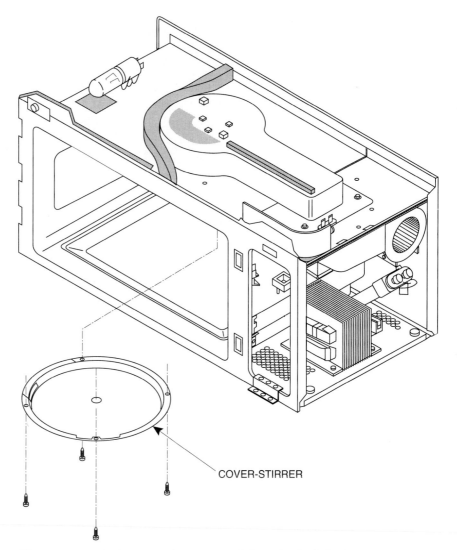

COVER-STIRRER

Figure 14-10.   A stirrer on a microwave oven is located under a plastic cover.

tern throughout the oven cavity. The stirrer assembly itself is often shaped like a propeller blade and can be driven directly by a motor mounted on the cavity, by a belt-driven system, or by an air-driven arrangement. The stirrer can be revealed by removal of the stirrer cover, such as the one shown in Figure 14-10.

14.3   LEAK TESTING MICROWAVE OVENS

The Federal Communications Commission recommends that microwave oven door seals be tested regularly to ensure there is no microwave leakage. Since the microwave oven operates similarly to a broadcast station, regulations regarding design and recommendations regarding safe operation were developed through the FCC. A *microwave oven survey meter* is a device that measures the

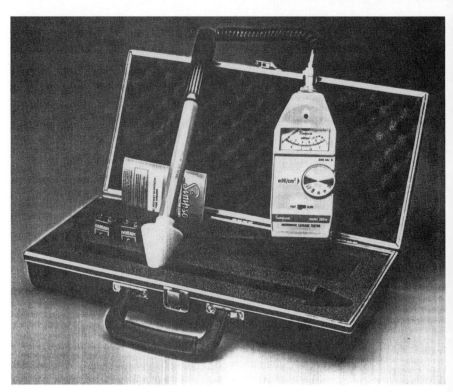

Figure 14-11. An approved, calibrated microwave oven survey meter is required to test microwave oven door assemblies for safety. *(Courtesy of Simpson Electric Co.)*.

amount of microwave leakage around a door seal. While leakage is a remote possibility due to the method of construction of the door and the design of the safety switch system, most customers will expect you to test their unit as part of the service provided in the process of completing the repair. A microwave survey meter is shown in Figure 14-11.

Microwave oven manufacturer guidelines will usually require that a leakage level of no more than 1MW/CM2 be allowed. Once a unit is in service, a reading of up to 5MW/CW2 is considered to be within safety limits.

The door assembly is constructed so as to prevent microwave leakage. A screen prevents microwaves from passing through the glass portion of the door assembly. Since a microwave signal is approximately as thick as a pencil point, manufacturers need only provide a screen in which the perforations are small enough in diameter to stop the microwave signal. Different types of seals are used around the door perimeter to make sure the appliance operates safely.

14.4   MICROWAVE OVEN WIRING DIAGRAM

As with all appliances, the schematic diagram of a microwave oven electrical system can give you an opportunity to identify its components and understand its method of operation. You'll notice, in Figure

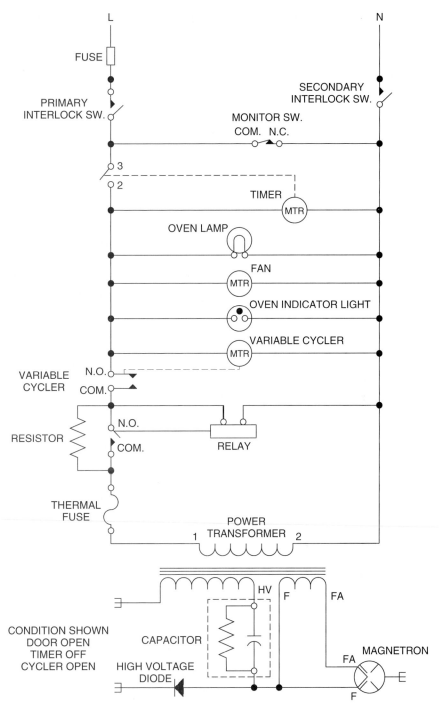

Figure 14-12. A schematic diagram for a microwave oven.

14-12, that the power transformer is shown as a load on the primary side and as a power supply on the secondary side. The 120-volt circuit is shown from L to terminal #1 on the primary side of the transformer and from N to terminal #2, providing the primary and secondary interlock switches are closed and the timer contacts between terminals #2 and #3 are closed.

The secondary windings of the transformer are shown in their relation to the voltage doubler circuit (the capacitor and high-voltage diode); the filament winding, noted FA, is also shown as a power supply to the magnetron tube. Other components necessary for the operation of the oven are also shown in the ladder diagram.

CHAPTER FOURTEEN   **SUMMARY**

Many customers are unnecessarily concerned about the safety of the microwave oven due to the fact that they don't understand the basic concepts behind its operation. Microwaves are electromagnetic waves of radiant energy that can be reflected, transmitted, or absorbed. They can be compared to the signal generated by a TV or radio station and to light waves. The frequency of operation of a microwave oven is 2,450 megahertz or 2,450 million cycles per second.

All food is made up of molecules, and the microwave cooks food by causing the molecules to vibrate back and forth in search of their proper pole. The vibration creates friction, which results in the heat that does the cooking. Microwaves penetrate to a depth of about 2 inches, and the inner layers of food are cooked through the heat transfer from the outer layer.

The *timer* on a microwave oven enables the customer to select a cooking time, and some models allow the selection of different cooking levels. The magnetron tube is the heart of the microwave oven, generating the microwave energy when the necessary voltage is provided by the transformer, capacitor, and diode.

The *stirrer assembly* helps to promote even cooking throughout the oven cavity. It may be belt-driven, driven directly by a motor, or air-driven. The cooling fan keeps the magnetron tube from overheating, and the door switches will not allow the unit to operate unless the door is properly closed.

Technicians are expected to conduct tests on the door to make sure there is no microwave leakage. An approved meter that has been calibrated within the prescribed time frame must be used.

CHAPTER FIFTEEN

Dishwashers

LEARNING OBJECTIVES   **After studying this Chapter, you will be able to:**
1. Describe the basic method of construction of a dishwasher.
2. Identify the electrical components in a dishwasher.
3. Describe the basic method of operation of different brands of dishwashers.
4. Trace a simple electrical circuit in a dishwasher.

■   ■   ■

Appliance technicians will often be required to install as well as service automatic dishwashers. The dishwasher *change-out*, replacing an old unit with a new one, is the most frequent type of installation. During a change-out, you'll be required to disconnect the power, water line, and drain line from the old machine, unfasten it from the underside of the countertop, then reconnect the water, drain, and electrical supply lines to the new unit, and fasten it to the countertop. Procedures in servicing specific models of dishwashers can vary as widely as those in regard to servicing laundry equipment. The first step in servicing them effectively is to establish an understanding of basic components and apply your basic troubleshooting skills, using common sense to solve the problems you find.

Construction methods vary. The most common method of motor mounting is center-mounted with the shaft pointing up into the tub. Seal assemblies are used to protect the motor. Some manufacturers use a side-mounted motor, and one manufacturer (Maytag) has in past years mounted the motor off to the side underneath the dishwasher tub and used a belt-driven system.

While some things are different, things such as water valves, timers, float switches, and door seals are the same in their basic function from one manufacturer to another. And that's true whether it's DM (Design and Manufacturing) Company who builds units under several different brand names such as Magic Chef, Admiral, some Sears units and others, Kitchen Aid, Maytag or Whirlpool. The basic method of operation—a wash cycle, a drain cycle, a rinse cycle, and another drain cycle—is the same.

15.1 UNDERCOUNTER DISHWASHERS

The *undercounter*, or *built-in, dishwasher* is the most common unit found in the home. Unlike the portable units, the construction of a built-in is accomplished without an outer cabinet. The tub assembly, which may be plastic or porcelain-on-steel construction, is mounted to a frame that supports it along with the door and electrical components. You may find a thin sheet of insulation around the outer surface of the tub. The purpose of this is to absorb sound. Some manufacturers accomplish sound absorption by spraying a foam insulation on the outer surface of the tub. Figure 15-1 shows a dishwasher tub as it is mounted to its support frame.

Leveling feet on the bottom of the support legs of the unit allow you to slide the unit under the counter, then raise the unit by adjusting them. Some models come with small plastic wheels built onto the back leg section. These allow you to wheel the dishwasher under the counter by lifting up on the front of the unit. Once you have the unit in place, adjusting the leveling legs will raise the unit up off the wheels.

All built-in dishwashers, regardless of manufacturer, are designed to fit into a standard cabinet opening. You'll find that some type of angle bracket is used to fasten the dishwasher cabinet to the underside of the countertop. Standard wood screws, short enough to prevent damage to the counter surface, are used to join the cabinet to

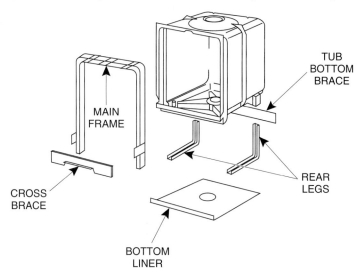

Figure 15-1. Dishwasher construction methods for both an undercounter and a portable dishwasher are fundamentally the same.

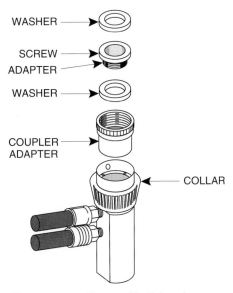

WASHER

SCREW
ADAPTER

WASHER

COUPLER
ADAPTER

COLLAR

Figure 15-2. The portable dishwasher coupler allows for connection to the kitchen faucet.

what is usually a particle board underside. (Dishwasher installation is covered in more detail in Appendix E.)

15.2 PORTABLE DISHWASHERS

You'll find portable dishwashers in older homes that cannot accommodate a built-in unit or in situations where your customer rents rather than owns the home. Electrical and mechanical construction of the portable unit is the same as the undercounter model, the exception being that the portable unit comes equipped with decorative cabinet panels on the sides and the rear. Portable units are equipped with wheels that allow the customer to move the unit near the sink and connect the water supply. A plastic connector, to which both the fill and drain hoses are connected, is attached to the kitchen faucet. Typically, an adaptor must be used on the faucet to allow for connection of the coupler unit. Figure 15-2 shows a portable dishwasher coupler.

The coupler allows for the water supply in the wash mode and also for drainage in the pump out mode. The hose assembly is stored in the dishwasher cabinet when not in use and is pulled out for connection to the faucet. The electrical cord is also stored in the cabinet on a spring-loaded retracting wheel.

15.3 DISHWASHER WATER SYSTEMS

Dishwasher water systems consist of the water supply, circulation, and discharge systems. A solenoid-operated water valve, similar to those found on a washing machine, is used to allow water flow into

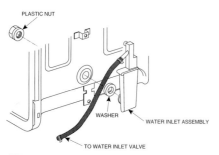

Figure 15-3. The water inlet assembly acts as an air break on a dishwasher.

the machine when the timer is in the fill mode. In most cases, a single solenoid is used. One manufacturer, KitchenAid, commonly uses a two-solenoid system. A two-solenoid water valve is not used to mix water temperature as it is on a washing machine. Both solenoids are positioned on the valve body so as to allow water flow into the machine on its one line when both coils are energized.

Water leaves the valve and travels along a rubber hose that is connected to a water inlet assembly. The water inlet assembly, such as the one shown in Figure 15-3, fastens to the side of the tub and acts as an air break, at the same time allowing water flow into the tub.

During the full cycle, the motor is energized by the timer, and the pump assembly attached to the motor accomplishes the second water system, the circulation during the wash and rinse cycles. The water in the bottom of the tub, usually 2 to 3 inches deep, is recirculated by the pump assembly during the wash mode and the rinse mode. A spray arm mounted on top of the pump assembly directs the water over the dishes. Some models may have more than one spray arm. Figure 15-4 shows one method of water circulation in a dishwasher.

Between the wash and rinse modes and after the final rinse, the dishwasher pumps the water out and down the drain. On some makes, this pump-out is accomplished by running the motor in a

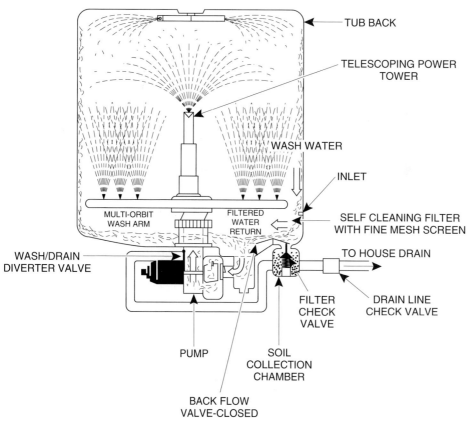

Figure 15-4. During the wash and rinse modes, the spray arm distributes water over the dishes.

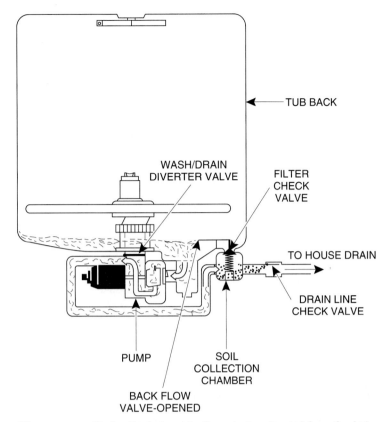

TUB BACK

WASH/DRAIN
DIVERTER VALVE

FILTER
CHECK
VALVE

TO HOUSE DRAIN

DRAIN LINE
CHECK VALVE

PUMP

SOIL
COLLECTION
CHAMBER

BACK FLOW
VALVE-OPENED

Figure 15-5.   During the drain mode, the water is redirected down the drain.

different direction than in the wash and rinse modes. Other manufacturers use a solenoid valve that, when energized, allows the water to exit the tub and be diverted down the drain rather than recirculated. In this case, the motor is designed to run in one direction only. Figure 15-5 shows a dishwasher in the pump-out mode.

15.4 DISHWASHER ELECTRICAL SYSTEM

Dishwashers operate on a standard 120-volt circuit, and most manufacturers will recommend that they be connected to their own 15- to 20-amp circuit. You will find some built-in units hard-wired, while others may have a power cord (attached by the technician upon installation of the unit) that plugs into a wall socket located in the cabinet underneath the sink. In either case, the unit must be properly grounded.

Other than the solenoid valve already discussed, other electrical components found in a dishwasher are the timer, door switch, float switch, detergent dispenser (either solenoid or bimetal), heating element, and motor and current relay. Figure 15-6 shows these components in a pictorial diagram.

The dishwasher *timer* operates similarly to the washing machine timer in that it energizes appropriate components throughout the

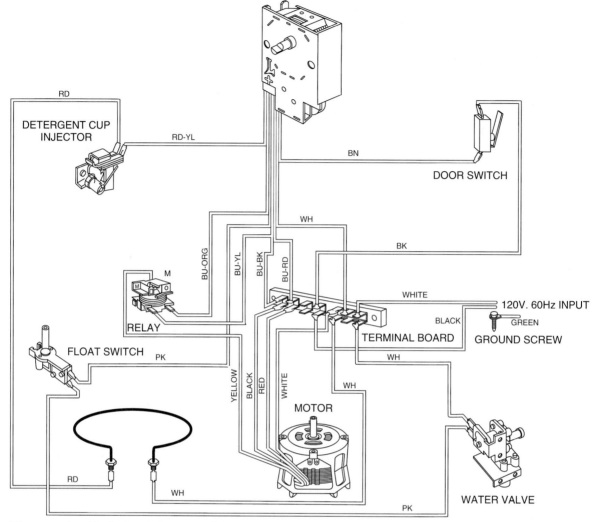

Figure 15-6. A basic electrical system and its components.

selected cycle. Some dishwasher timers are equipped with two motors: the standard motor that allows the cycle selected to run its course on normal time frame, and a rapid advance motor that drives the timer through its cycles very quickly. This feature allows the customer to cancel the cycle and drain all the water from the machine.

The *door switch* on the dishwasher is wired in such a way as to prevent any operation of the unit at all if the switch is not in the closed position. The switch is mounted so that it is contacted by the latching assembly on the door. This is a standard single-pole, single-throw switch.

The *float switch* is a small switch located under the tub. Wired in series with the water valve, it is used to break the circuit to the water valve and prevent overflow in the event that a malfunction with the valve or the timer allows water to continue flowing into the tub. Figure 15-7 shows the method of operation of the float switch.

The *plastic float* and accompanying shaft is heavy enough to keep the switch in the closed position when there is no water in the

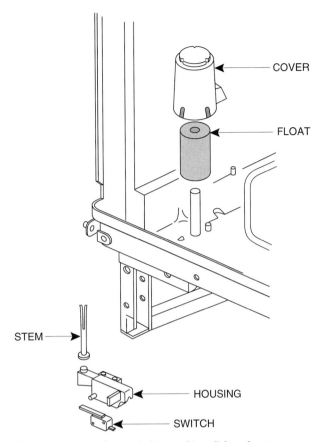

Figure 15-7. A float switch is used in a dishwasher to provide the correct level of water in the tub.

tub. When the tub fills with water, the float rises and subsequently the switch opens, breaking the circuit to the water valve.

The *detergent cup dispenser* may be a solenoid-operated device, but is more commonly found to be a bimetal unit. When you close the door on the detergent dispenser assembly, you're actually locking the door in place because the bimetal is unaffected by current flow at this time and allows the door to lock.

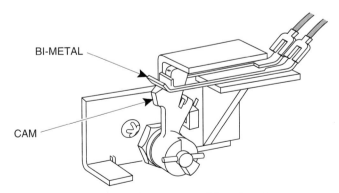

Figure 15-8. The soap dispenser on a dishwasher often ues a bimetal device to trip the soap dish during the wash mode. When the customer puts soap in the dispenser and closes the door, it locks into place on the bimetal.

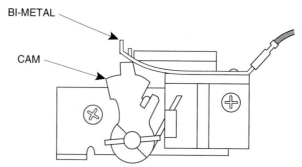

Figure 15-9. As power is applied through the bimetal device, it warps and releases the soap dispenser door.

The bimetal will, on some older units, be wired in series with the motor winding. So, while the tub is not yet full of water, the current draw of the motor is low and the bimetal will be unaffected

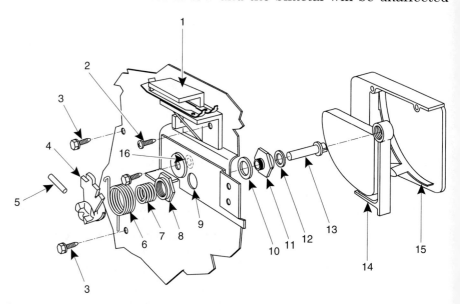

Reference Number	Description
1	Bi-Metal Trigger
2	Screw
3	Screw
4	Cam
5	Rollpin-Cam
6	Spring-Cam
7	Spring Cam-Compression
8	Nut
9	Bracket-Actuator
10	Washer
11	Bushing
12	"O" Ring
13	Shaft
14	Cup-Detergent
15	Cover-Cup
16	"O" Ring

Figure 15-10.

by the current flow through it. As the tub fills with water, the motor has to work harder, and as a result the current draw of the motor increases. The increase affects the bimetal device, causing it to warp and release the dispenser. Figure 15-8 shows the dispenser in a locked position, and Figure 15-9 shows the bimetal warping and allowing the release of the detergent door. Figure 15-10 shows the complete detergent dispensing assembly, as it is shown in a manufacturer's service manual.

Most dishwashers are equipped with a *heating element* such as the one shown in the bottom left corner of Figure 15-6. This element is similar in construction to the bake elements and defrost heaters already discussed. In a unit without a forced-air drying system (which you'll find on most "top-of-the-line" dishwashers), this heating element, depending on the make and model, can serve two purposes: first, to make sure that the water in the dishwasher reaches the minimum 140° Fahrenheit recommended by many manufacturers, and, second, to provide radiant heat in the drying mode. The element is positioned in the dishwasher bottom in such a way as not to create a leak, and the wiring that provides energy to the element is routed underneath the dishwasher tub. In some cases, this type of heating element will be wired in series with a bimetal thermostat that is positioned in a clip close to the dishwasher tub. The thermostat is designed to cycle the heating element off when 140° Fahrenheit is attained. During the wash and rinse cycles, the heater is immersed in water.

On a unit not equipped with a 140° cycling thermostat, the heating element will be energized only in the sanitize cycle if the machine is equipped with this feature. In this case the heating element will serve to raise the water temperature to 150–180° Fahrenheit depending on the manufacturer's design.

Dishwasher *motors* are designed and manufactured in many different styles. The most popular configuration, though, is the center-mounted, shaft-up system. With this method of design, the motor is mounted in a housing that allows it to hang underneath the dishwasher in an exactly centered position. A series of seals and impellers is used to protect the motor from leaks and accomplish the washing and rinsing processes.

Figure 15-11 shows one manufacturer's motor as illustrated in a service manual. As you can see, the pump/motor housing contains the various gaskets, seals, plates, and even a food chopper that make up the water circulation and pump out system. The type of system shown uses a motor that runs in one direction for the wash mode and in another direction for the rinse mode. This means that this motor has two start windings to enable it to run both clockwise and counterclockwise. Different contact points in the timer accomplish these circuits as the cycle runs in course. A current relay is used to momentarily energize the start winding of the motor. The current relay operates on the same principle as the one we've discussed in the unit

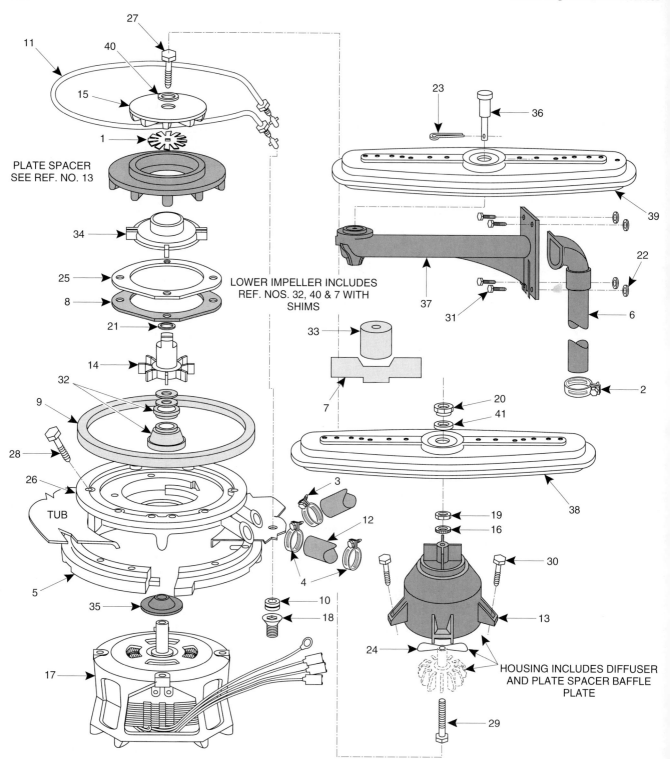

PLATE SPACER
SEE REF. NO. 13

LOWER IMPELLER INCLUDES
REF. NOS. 32, 40 & 7 WITH
SHIMS

TUB

HOUSING INCLUDES DIFFUSER
AND PLATE SPACER BAFFLE
PLATE

Figure 15-11.

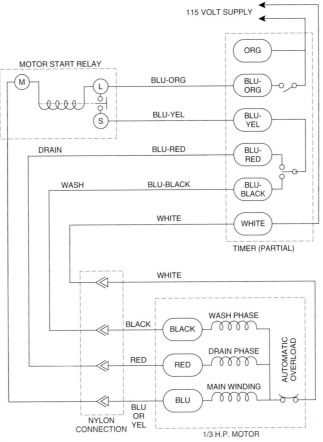

Figure 15-12.

on refrigerators and freezers, the difference being that there are no pin terminals to connect to on the dishwasher motor. The relay in this case is simply connected with a wire harness.

A simplified wiring diagram for a clockwise/counter clockwise motor is shown in Figure 15-12. Tracing the start/run circuit with this diagram is a fairly simple task. Start with the 120-volt power supply at the top of the diagram and follow the hot wire (labeled BLU-ORG, usually meaning a blue wire with an orange tracer), and follow that wire as it connects to L on the motor start relay. This hot wire goes on through the coil on the motor start relay, leaves on M, and then runs directly to the main winding on the motor. As you continue to trace the circuit through the main winding, you'll see the motor protector, labeled in this case as the automatic overload, and you can trace the white wire from the main winding in an unbroken path back to the neutral side of the 120-volt power supply.

The second circuit for you to consider in the starting and running of this motor is through one of the start windings. For our example, we'll trace the circuit through the wash phase. Starting again with the hot leg through BLU-ORG to L on the relay, follow it through the S terminal on the relay and begin the circuit through the BLU-YEL wire. Remember, a current relay's contacts are normally in an open position until the current draw of the run winding causes them to close momentarily. That means, at the instant of start, the path of electrical flow is not only through the relay coil as already described, but also through that normally closed switch to the appropriate start winding. In our example, this circuit goes through the BLU-YEL wire to the timer, then through the switching contacts inside the timer. It leaves the timer on BLU-BLACK and connects to the wash phase start winding. And, as with the main winding circuit, the neutral side of the wash phase start winding circuit is provided through the white wire back to the 120-volt supply.

At the instant of start, both the main and start windings are powered. As the motor picks up speed (this happens very quickly), the current draw in the main winding goes up. As a result, the magnetic field set up in the coil of the relay is stronger—strong enough to overcome the gravity and spring pressure holding the rod that provides a method for jumping across the L and S contacts. The gravity and spring pressure are overcome only briefly, and, as the motor picks up speed, the current draw of the run winding drops and the points inside the relay are separated. This leaves only the circuit we

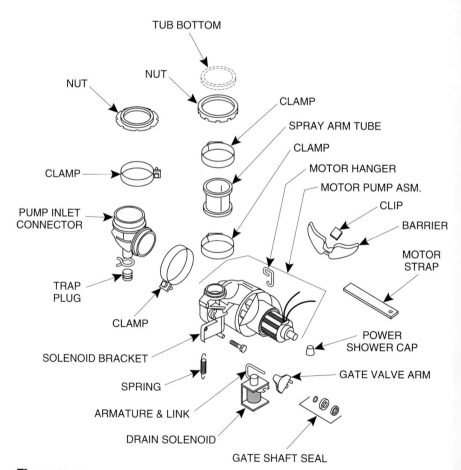

Figure 15-13.

first traced through the main winding in place and the motor continues to run.

As the timer advances through the wash mode, the motor is stopped, then restarted in the drain mode through different contacts of the timer. The circuit through the drain phase winding can be traced from BLU-RED on the timer.

Not all dishwashers operate in the manner described. Another example would be the side-mounted motor that does not reverse in direction. Instead, it uses a solenoid system to accomplish the draining of the wash and rinse water as the timer goes through its cycle. A motor setup of this type is shown in Figure 15-13.

CHAPTER FIFTEEN   **SUMMARY**

Appliance technicians are often called on to remove and install new dishwashers, as well as to troubleshoot problems and replace parts. Two kinds of dishwashers found in the home are the built-in unit and the portable. *Built-in units*, regardless of the manufacturer, will fit into a standard cabinet opening. *Portable units* are used by customers in apartments or in homes in which the cabinet work cannot accommodate an undercounter unit.

Portable and undercounter units are constructed in fundamentally the same manner, the difference being that the portable unit has decorative cabinet panels. You may also find that the insulation (either fiberglass or foam sprayed onto the outer surface of the dishwasher tub) is used to deaden the running noises of the dishwasher. On a portable dishwasher, the faucet coupler assembly allows for water supply connection and also for the drain hose connection.

On a dishwasher, the two water systems to consider are the water supply and water circulation systems. The *water supply system* uses a solenoid valve, which, when energized, allows water to flow into the dishwasher tub through a water inlet assembly. Some manufacturers use two solenoids on the water valve as a safety measure. Should one section of the valve fail in an open position and create a flooding situation, the other solenoid would close and prevent the overflow.

During *water circulation*, the spray arm distributes the water over the dishes in both the wash and rinse modes. During the drain mode, the pump assembly redirects the water flow to the drain instead of through the spray arm.

The standard electrical components found in a dishwasher are the motor, water valve, door switch, timer, detergent cup injector, float switch, and heating element. Some dishwashers may also be equipped with a forced-air blower system. One popular method of operating a dishwasher is with a motor that runs in one direction for the wash mode and in another direction for the rinse mode. A current relay is commonly used to start of the motor on a dishwasher.

Sealed System Servicing

As an appliance technician, you will spend a relatively small percentage of your time servicing the sealed system on refrigerators and freezers. While most of your time will be spent solving electrical problems on various types of appliances and performing mechanical procedures such as replacing pumps and motors, understanding refrigeration systems and knowing proper service procedures is a critical part of your job. It's common for inexperienced technicians to rush into working with the sealed system on a unit that is reported as "running but not cooling" when in fact such a symptom could be indicative of an electrical or air circulation problem. With that thought in mind, we'll go through a short review of the fundamentals of refrigeration as covered in Chapter Six, while at the same time laying the foundation for the development of your troubleshooting skills in regard to refrigeration systems. Refer to Figure A-1.

Compressor. This is a vapor pump that accepts a low-pressure vapor and discharges it as a high-pressure vapor. A compressor may be of the reciprocating (piston) design or a rotary type. Horsepower ranges from a low of $1/12$ for smaller units to a high of $1/3$ for larger units such as side-by-sides or top mount refrigerators above 16 cubic feet. (A bottom freezer of 16 cubic feet and above would also use a $1/3$ HP compressor.)

A hermetically sealed compressor contains an electric motor assembly, as well as the components that accomplish the compression of the refrigerant vapor. Other than an electrical failure, a compressor can fail mechanically by seizing up, in which case it would hum but not start, or there may be a valve failure. In this case, the

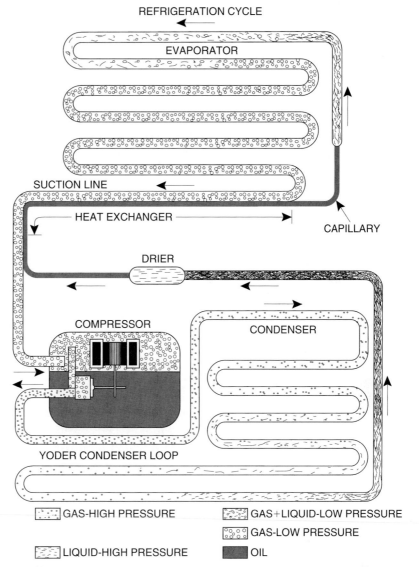

Figure A-1. A basic refrigerator sealed system. The components, direction of refrigerant flow, and state of refrigerant are shown.

compressor would run but it would not pump efficiently. In other words, a valve failure would prevent a compressor from discharging at the required high pressure, and the result would be that the design temperature of the unit would not be achieved. Some cooling would be accomplished, but, depending on the severity of the valve failure, the operating temperatures of the refrigerator could be 10, 15, and even 20° higher than normal. This would result in the customer complaining of soft ice cream in the freezer, an automatic ice maker not working, and accelerated food and milk spoilage in the fresh food compartment.

Condenser. The condenser is located on the high-pressure discharge side of the compressor. Its function is to transfer the heat absorbed by the refrigerant into the ambient (room) air. High-pressure gas is routed to the condenser, where, as the gas temperature is reduced, it condenses into a high-pressure liquid. The heat transfer takes place because the discharged gas is at a higher temperature than the air passing over the condenser tubing.

A condenser can be made of tube and wire, and mounted on the back of the refrigerator cabinet (this is referred to as a *static condenser* by most manufacturers). Or it can be fan-cooled type. Failure of the condenser fan motor would cause higher-than-normal operating temperatures because the efficiency of heat transfer from the condenser would be affected by the impair air flow. This would be one example of a "running but not cooling" complaint that would *not* be caused by a sealed system malfunction.

A condenser on the simplest of refrigeration systems would be a single section. On more sophisticated units, the condenser may contain a loop of tubing that serves to keep the refrigerator cabinet from sweating. Or a portion of the condenser may serve as an oil cooling loop that is routed (as a closed loop of tubing) back into the compressor and out again before forming the main section of the condenser. In some cases, manufacturers will position the drain pan on a section of the condenser to use some of the heat to help evaporate the defrost water.

A leak in the condenser tubing can usually be detected by use of an electronic leak detector or a soap solution. Since oil is always circulating with the refrigerant throughout the sealed system, a telltale sign of a condenser leak would be a small amount of oil on the tubing near the leak.

Drier. The filter/drier on a refrigeration system is placed at the end of the condenser. Since the change of state from a gas to a liquid takes place in the condenser, the filtering material inside the filter/drier is designed to remove contaminants from a liquid. The filter/drier is constructed in a manner that allows attachment to the condenser (usually $\frac{1}{4}$- or $\frac{3}{16}$-inch tubing) on one end and accepts the capillary tube in a much smaller opening at the other end. Most filter/driers are directional with an arrow indicating the direction of refrigerant flow but some may be installed in either direction. The filter-drier must be changed whenever the sealed system is opened for any reason, such as a compressor replacement or leak repair. If a sealed system is restricted, replacing the filter/drier can sometimes solve the problem.

Capillary tube. The capillary tube is the metering device in a domestic refrigeration system and is sized in diameter and length to feed the proper amount of refrigerant to the evaporator. It is soldered to the suction line (or, in some cases may be routed inside the suction line) to transfer heat from the warm capillary tube to the cooler suction line. This forms the heat exchanger, the function of which is to add additional superheat to the refrigerant gas in the suction line and prevent the migration of liquid refrigerant to the compressor. Another function of the heat exchanger is to ensure that the liquid refrigerant that discharges from the capillary tube into the evaporator is at the correct temperature, enabling it to do maximum work in the evaporator. A separation of the heat exchanger (the solder bond between the suction line and the capillary tube breaking down) will affect the efficiency of the refrigeration system operation since the refrigerant will not be entering the evaporator at optimum temperature.

Evaporator. The inner volume of the evaporator, since it is much larger than the inner volume of the capillary tube, allows the liquid refrigerant discharged from the "cap tube" to expand into a refrigerant gas. A change in state from a liquid to a vapor allows for the absorption of heat from the air that is either being forced through the evaporator or that circulates naturally across it. Some manufacturers refer to the capillary tube as an expansion device rather than as a metering device because expansion takes place inside the evaporator and allows the evaporator assembly to be cooled to a approximately –20° Fahrenheit. The refrigerant exits the evaporator and is pulled through the suction line to the compressor, where the refrigerant cycle begins again.

As mentioned, there are many different design configurations for the sealed system of a refrigerator/freezer or freezer. A variety of methods are used to accomplish the refrigeration process, and condensers and evaporators may take many different forms.

Figures A-2 and A-3 are just two more representations of the refrigeration system just reviewed. Figure A-2 is an illustration from an Amana service manual, and Figure A-3 is an illustration from a Maytag (Maycor Corp.) refrigerator service manual. Both units use fan-cooled condensers, but, as you can see, they are different in appearance. The Maytag unit uses a pass of the condenser tubing to form what is known as a *yoder loop*. This allows for an energy-saving method (using some of the heat from the condenser that had to be dispensed with anyway) to keep the cabinet area around the freezer from sweating. An alternative method would be to use an electric heater, which uses energy.

The Amana unit uses one section of the condenser to aid in the evaporation of the defrost water. You'll also notice that an accumulator is shown in Figure A-3. This is a sealed system component that is located at the outlet of the evaporator. It is nothing more than a storage tank. Its function is to ensure that liquid refrigerant does not get to the compressor.

No matter what the configuration or manufacturer, certain troubleshooting and diagnosis procedures apply to sealed system servicing. Just keep in mind that, if a system is operating properly, it is operating within given temperatures and pressures. To know if a system is malfunctioning, you first must be familiar with what you would find if you

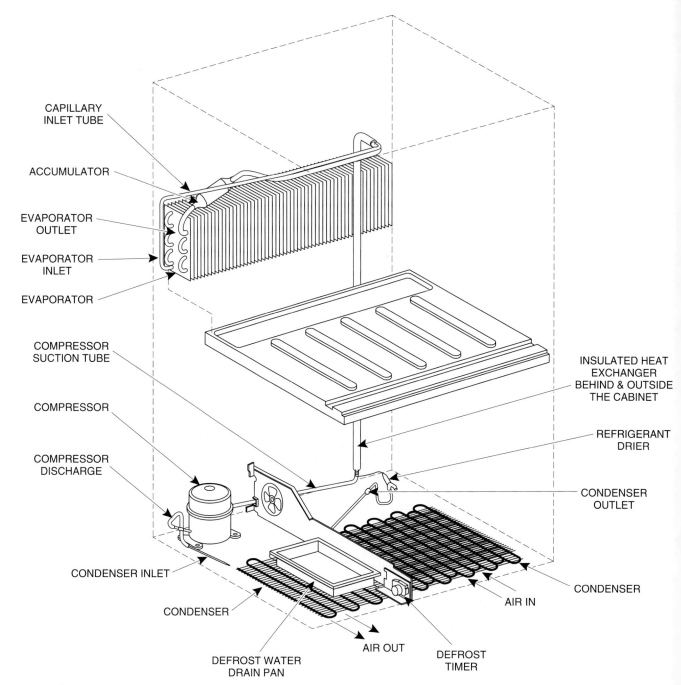

Figure A-2. An illustration of the sealed system of one model offered by Amana. *(Drawing courtesy Amana Refrigeration, Inc. Amana, Iowa).*

checked a system that is operating normally. For our purposes, we'll establish that our normally operating system is a standard R-12 unit that is sitting in a room in which the ambient temperature is 75° Fahrenheit. The room temperature is a factor because, since there is a temperature/ pressure relationship between refrigerants, a change in room temperature, has an effect on high-side pressure readings: the higher the room temperature, the higher the pressure. As the temperature goes down,

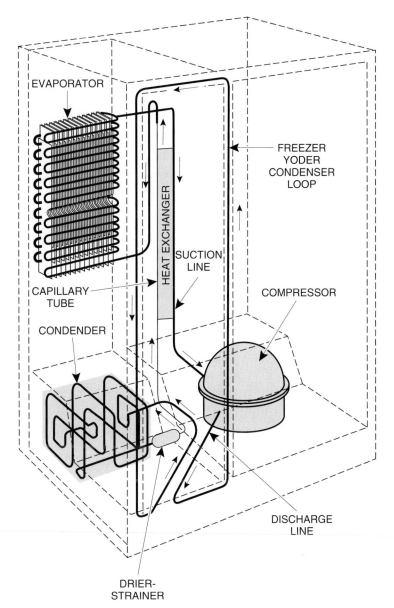

Figure A-3.   An illustration of the sealed system of one model of a Maytag refrigerator. *(Courtesy Maycor Corp.).*

the pressure goes down. the low-side pressure reading on a normally operating refrigerator that is already cooling and has cycled on and off several times is fundamentally not affected by the ambient temperature.

For our discussion on sealed system troubleshooting, we'll be covering the following specific problems:

System overcharge

Undercharge (no leaks, incorrect factory or field charge)

Undercharge (leak on the low side)

Undercharge (leak on the high side)

Partial restriction (high side)

Partial restriction (low side)

Complete restriction (due to a solid contaminant)

Complete restriction (due to moisture in the system)

Inefficient compressor (bad valves)

Air in system

FREON®
REFRIGERANTS

Temp. (°F)	Vapor Pressure*				
	FREON 11	FREON 12	FREON 22	FREON 500	FREON 502
−150	29.9	29.6	29.4		29.1
−140	29.9	29.4	29.0		28.5
−130	29.9	29.1	28.4		27.8
−120	29.9	28.6	27.7		26.7
−110	29.8	28.0	26.6		25.3
−100	29.8	27.0	25.0	26.4	23.3
−90	29.7	25.7	23.0	24.9	20.6
−80	29.6	24.1	20.2	22.9	17.2
−70	29.4	21.8	16.6	20.3	12.8
−60	29.2	19.0	12.0	17.0	7.2
−50	28.9	15.4	6.2	12.8	.2
−40	28.4	11.0	.5	7.6	4.1
−35	28.1	8.4	2.6	4.6	6.5
−30	27.8	5.4	4.9	1.2	9.2
−25	27.4	2.3	7.4	1.2	12.1
−20	27.0	.6	10.2	3.2	15.3
−15	26.5	2.4	13.2	5.4	18.8
−10	26.0	4.4	16.4	7.8	22.6
−5	25.4	6.7	20.1	10.4	26.7
−0	24.7	9.2	24.0	13.3	31.1
5	23.9	11.8	28.2	16.4	35.9
10	23.1	14.6	32.8	19.7	41.0
15	22.1	17.7	37.7	23.3	46.5
20	21.1	21.0	43.0	27.2	52.4
25	19.9	24.6	48.8	31.4	58.8
30	18.6	28.4	54.9	36.0	65.6
35	17.2	32.6	61.4	40.8	72.8
40	15.6	37.0	68.5	46.0	80.5
45	13.9	41.7	76.0	51.6	88.7
50	12.0	46.7	84.0	57.5	97.4
55	10.0	52.1	92.6	63.9	106.6
60	7.8	57.7	101.6	70.6	116.4
65	5.4	63.8	111.2	77.8	126.7
70	2.8	70.2	121.4	85.3	137.6
75	0.0	77.0	132.2	93.4	149.1
80	1.5	84.2	143.6	101.9	161.2
85	3.2	91.8	155.7	111.0	174.0
90	4.9	99.8	168.4	120.4	187.4
95	6.8	108.2	181.8	130.5	201.4
100	8.8	117.2	195.9	141.1	216.2
105	10.9	126.6	210.7	152.2	231.7
110	13.1	136.4	226.4	164.0	247.9
115	15.6	146.8	242.7	176.3	264.9
120	18.3	157.7	259.9	189.2	282.7
125	21.0	169.1	277.9	202.8	301.4
130	24.0	181.0	296.8	217.0	320.8
135	27.1	193.5	316.6	231.8	341.2
140	30.4	206.6	337.2	247.4	362.6
145	34.0	220.3	358.9	263.7	385.0
150	37.7	234.6	381.5	280.7	408.4

*Vapor pressures are shown as PSIG.

Chemical Names and Formulas for FREON Refrigerants

Registered Trademark	Chemical Name	Formula	Boiling Point °F
FREON 11	Trichlorfluoromethane	CCl_3F	74.9
FREON 12	Dichlorodifluoromethane	CCl_2F_2	−21.6
FREON 22	Chlordifluoromethane	$CHClF_2$	−41.4
FREON 500	Azeotrope of FREON 12 and 1,1-difluorethane		−28.3
FREON 502	Azeotrope of FREON 22 and FREON 115		−49.8

Figure A-4. A temperature/pressure chart is used to evaluate a refrigeration system.

For comparison, our normally operating unit factors are as follows:

Low-side operating pressure: 0-5 pounds

High-side operating pressure: (75° ambient) 126.6 pounds

Before we proceed with information on gaining access to a sealed system and the use of gauges to diagnose a problem, it's important that you understand how we arrived at the preceding information. The explanation centers around what is known as a temperature/pressure chart, such as the one shown in Figure A-4. A *TP chart* illustrates the concept of the temperature/pressure relationship between refrigerants. The two columns we'll be concerned with are the "Temp" column at the far left of the chart and the column marked "Freon 12." Freon 12 is a registered brand name of the DuPont corporation for refrigerant 12. The other refrigerants shown on the chart (Freon 11, Freon 22, Freon 500, and Freon 502) are used in systems other than refrigerators and freezers, and we won't be discussing their application. Positive pressures are in black, negative pressures (operation in a vacuum) in gray.

To understand how we arrived at the operating pressures for our normally operating system, begin by looking at the temperature column at the left of the chart. You'll recall that we said our evaporator temperature will be near –20° Fahrenheit. To understand why our normal low-side operating pressure is near 0 pounds, go to –20° on the chart, then read over to the Freon 12 column. You'll see that the corresponding vapor pressure in the Freon 12 column is 0.6 pounds. Most manufacturers will tell you that the rule of thumb to apply is that a reading between 0 and 5 pounds pressure, depending on whether some warm air has been introduced by opening the cabinet door, would mean that the unit is operating properly. If you're checking a system and are able to monitor it for a reasonable length of time without the introduction of some warm air into the cabinet, your normally operating unit will read closer to 0 pounds.

To understand how we arrived at the normal high-side operating pressure, go again to the temperature column and read down to 75° Fahrenheit, which is what we decided the ambient temperature in our room would be. Another rule of thumb applies. To properly calculate the high-side operating pressure of an R-12 system operating in a 75° ambient, add 30° and read down to 105°. The reason the 30° must be added is to allow for heat added due to the operation of the system's compressor. The mechanical components of the compressor, such as the crankshaft and piston rods, are responsible for friction, and a factor known as the *heat of compression* (related to the compression of the refrigerant) are the reasons the 30° figure must be added to the ambient temperature.

Reading across from 105° to the Freon 12 column yields a pressure reading of 126.6 pounds, the high-side operating pressure listed on our properly operating unit. If you connect a set of gauges to a unit operating in a 75° ambient and you read numbers different than those that we've calculated, there is a problem either with the sealed system itself or with another factor, such as air flow over the condenser or evaporator. The possibility of air flow problems being a factor can be eliminated by visual inspection before tapping into the sealed system. You can remove a cover on an evaporator, for example, and see if the coil is clogged with frost or if the fan motor is not running. You can inspect a fan-cooled condenser for dust, dirt, or pet hair, eliminating that as a possibility beforehand.

A.1 GAINING ACCESS TO A SEALED SYSTEM

Figure A-5. A temporary piercing valve that allows access to the sealed system.

Once you have determined that air flow or electrical problems are not responsible for improper operation of the refrigerator, you will have to proceed with the testing of the sealed system. Since refrigerators do not come equipped with access valves that allow for the connection of gauges, you'll have to install them. (If you come upon a system with access valve on it, somebody's been there ahead of you.) Fundamentally, there are two types of access valves. One style, soldered onto a system, is considered permanent and accepted by manufacturers as necessary when performing warranty repairs. The other is a piercing type of valve that is considered temporary, used only for diagnosis and the recovery of refrigerant from a system. Manufacturers have been known to void a warranty on a unit on which the service technician has left a tap line valve behind after performing a repair. Many tap line valves are known to leak after a period of time.

There are many different styles of tap line valves. Fundamentally, they are a saddle-type valve that is secured to the refrigeration system tubing through two or three screws or by a saddle and nut

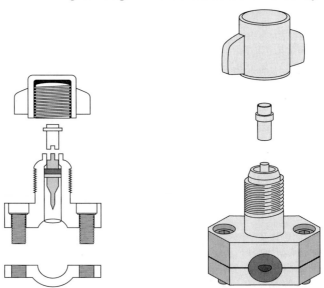

Figure A-6. A type of piercing valve that uses a spring-loaded valve core.

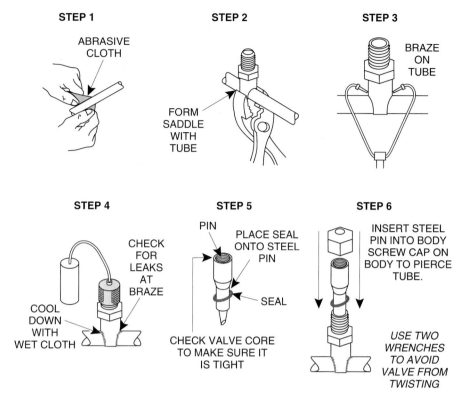

Figure A-7. Installing a permanent access valve.

assembly. Some are made to fit one size of tubing only, while others may use a spacer to allow installation on two or three different tubing sizes. They may come equipped with a handle that allows them to be opened or closed, such as the type of valve shown in Figure A-5. Or they may be more sophisticated, requiring a needle depressor to pierce the line such as the type shown in Figure A-6. Whatever their nature, they all achieve the same objectives: to allow the connection of a charging hose and testing and servicing of the sealed system.

Another style of access valve is identified as permanent and is soldered onto the refrigeration tubing. This type of valve is made up of a copper and brass fitting that is attached to the tubing and soldered in place; a core assembly is added to complete the installation. Figure A-7 shows the steps to follow in installing this type of access valve.

A.2 PERMANENT ACCESS VALVE INSTALLATION PROCEDURES

Step 1: Make sure the tubing is free of dirt and grime by cleaning the surface with an abrasive cloth.

Step 2: Position the valve body on the tube, and use a pair of pliers to squeeze the tails of the body together to form a saddle around the tube.

Step 3: Using a suitable torch, braze the valve body onto the tube. Use silver solder. (The melting temperature of the solder will be near 1,100° Fahrenheit.)

Step 4: Check for leaks by attaching a hose from a tank of dry nitrogen. Test with a soap solution after the valve body and tubing are cooled down.

Step 5: Place the nylon seal onto the steel piercing pin assembly. The valve core assembly, which closely resembles the valve core found in a bicycle inner tube and must be removed and reinstalled with a special tool, may stay in place in the pin assembly.

Step 6: Insert the piercing assembly into the valve body, and use the valve cap to force the piercing assembly down into position where it will pierce the tubing. Remove the cap after piercing the tubing. The valve assembly is now ready to accept a charging hose equipped with a valve depressor assembly that will press down on the spring-loaded valve core assembly.

Caution! *Permanent access valves that must be soldered onto a system should only be installed after the refrigerant in a sealed system has been removed. To facilitate recovery of the refrigerant and system diagnosis, use a saddle type valve for temporary access.*

Two other types of access valves are popularly used on refrigerators. One style, a threaded brass fitting soldered to a short length of copper tube, allows for the insertion either into a tee installed on a refrigerant line or directly into a process stub on the compressor. Another type of access valve is available already installed on the replacement filter/drier, such as the one shown in Figure A-8. This type of drier allows for ready-made access to the high-pressure side of the system in the process of replacing the system's filter/drier.

A.3 USING COMPOUND GAUGES

Once access to the sealed system has been gained, you can use your gauges to test, diagnose, and service the sealed system. The most common type of gauge set used is the three-hose type shown in Figure A-9. They are referred to as a *compound gauge set* because the gauge on the left reads pressure on the low side of the system and the gauge on the right reads pressure on the high side of the system. The hoses are equipped with valve core depressors at one end to allow for depression of the spring-loaded valve assembly.

Figure A-8.   A filter/drier with a built-in access valve.

Figure A-9. A three-hose manifold set. *(Courtesy Robinair).*

With the advent of refrigerant recovery systems, manifold gauges with four hoses are becoming more popular. Equipment manufacturers also offer gauges that give you a digital readout rather than using a needle. There are also oil- and glycerine-filled gauges on the market. The philosophy behind an oil-filled gauge is to dampen needle vibration and protect the gauges against any sudden pressure surge. Some gauge sets also come equipped with sight glasses that allow to you to monitor the charge to see if gas or liquid is being transferred into the refrigeration system.

When you are installing charging hoses on an access valve, always be sure that both sides of the manifold assembly are in the closed position. The manifold gauge assembly valves close clockwise and open counterclockwise. With the manifold assembly facing you, this would translate to turning the knob on the right away from you and the knob on the left toward you, to make sure both valves are in the closed position.

When using the manifold gauge assembly to test system operating pressures, leave the center hose attached to its threaded stub on the front of the manifold assembly. This hose is traditionally used only for connection of a refrigerant recovery system or vacuum pump, or to allow for adding refrigerant into the system.

To better understand the high- and low-pressure gauges, refer to Figure A-10 which offers a more detailed view of the markings and

Figure A-10. The high- and low-side guages allow an appliance technician to evaluate the operation of a refrigeration system. *(Courtesy Robinair).*

indicators on them. There are several scales on each gauge. The outermost scale represents pressure, and, as you can see, the high-pressure gauge on the right will show pressure from 0 to 600 PSI. The low-side gauge on the left will show positive pressure from 0 to 350 PSI, and will also read a negative pressure from 0 down to 30 inches of vacuum. The inner scales on each gauge are identified as indicating temperature in degrees Fahrenheit, and there is a scale for three different refrigerants (R-22, R-12 and R-502). For our purposes, we'll focus on the R-12 scale only. The purpose of these temperature scales is to show you what the temperature of the refrigerant will be when the needle indicates a given pressure. The R-12 temperature scale reads from a low of –40° Fahrenheit to a high of 100° Fahrenheit and is represented by the center row of the three temperature scales.

You'll note that the position of the needle on the low-side gauge indicated 0 PSI. Referring the R-12 temperature scale, you can see that this pressure corresponds closely to –20° Fahrenheit. You'll recall that in our example of a normally operating refrigerator, the evaporator coil was chilled to approximately –20°, and we said that the system low-side pressure would read near 0 PSI. If we were chilling the coil to near 0°, this would correspond to 10 PSI on the gauge.

Referring to the high-side gauge, if the needle were indicating 126.6 PSI, you can see that the corresponding temperature on the R-12 scale would be near 105°, the temperature we calculated on the temperature/pressure chart by adding 30° to the ambient temperature.

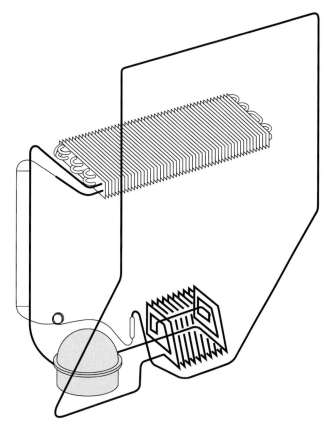

Figure A-11.   A simplified configuration of a standard domestic refrigeration system.

A.4   TROUBLESHOOTING SEALED SYSTEM PROBLEMS

Now that you know how to install access valves and gauges, we'll look at several systems that are not operating properly and the conditions you'll find for each of these problems. The refrigeration system we've chosen is a frost-free unit with a jelly roll condenser, such as the one described in Chapter Seven, but the conditions and symptoms described in each of the individual problems shown could apply to other refrigeration systems.

Before proceeding with the system malfunctions, make sure you're familiar with the system as it is shown in Figure A-11. The components shown are the compressor, condenser (as an assembly next to the compressor and in the form of a yoder loop), filter/drier, capillary tube, evaporator, and suction line. To make sure you understand the system, trace the refrigerant flow through the system's components. You'll note that the discharge line from the compressor goes directly into the condenser assembly. The refrigerant flow is then through the yoder loop, on into the filter/drier, then through the capillary tube before going through the evaporator and down the suction line to complete the refrigerant circuit back to the compressor.

Once you're familiar with the system, go ahead to the drawings and descriptions detailing the various system problems. In addition

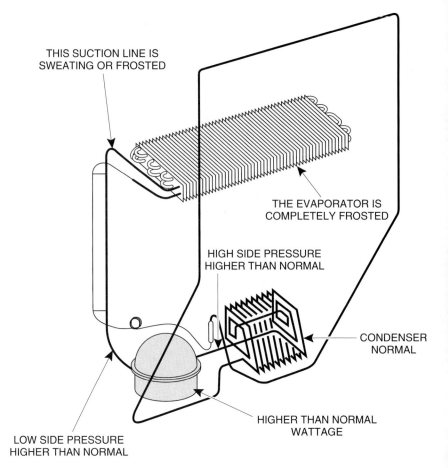

THIS SUCTION LINE IS
SWEATING OR FROSTED

THE EVAPORATOR IS
COMPLETELY FROSTED

HIGH SIDE PRESSURE
HIGHER THAN NORMAL

CONDENSER
NORMAL

HIGHER THAN NORMAL
WATTAGE

LOW SIDE PRESSURE
HIGHER THAN NORMAL

OVERCHARGE

Figure A-12. An illustration of an overcharged system.

to indicating whether system pressures are higher or lower than normal, we'll also be indicating what you may feel and hear when checking each one of the systems. You'll also note that one electrical
check—whether or not the wattage draw of the unit is higher or
lower than normal—is being used. You'll recall that we discussed the
use of a wattmeter in Chapter Five.

Caution! *Appropriate safety glasses should be used whenever
you are working with refrigerants and refrigeration systems. Liquid refrigerants can cause severe frostbite burns.*

Figure A-12 shows an overcharged system. Both high- and low-
side pressures are higher than normal, the frost pattern on the evaporator will be complete, and the suction line will be sweating or
possibly have a coating of frost. The wattage is higher than normal
because there is too much refrigerant in the system and the compressor is doing more work than it is designed to do. One approach to
solving this problem would be to install your gauges, recover all the
refrigerant in the system, then use one of the several different types
of measuring devices designed to help you add the correct charge to
a refrigeration system.

There are various types of refrigerant recovery systems on the market, and each is used according to the equipment manufacturer's specific instructions. One type of recovery system is shown in Figure A-13.

To measure the correct charge into a system, a charging cylinder may be used. Charging scales that monitor refrigerant flow electronically are also available. A charging cylinder is shown in Figure A-14, and an electronic scale is shown in Figure A-15.

The charging cylinder and the scale both allow you to charge a system using liquid refrigerant, the most effective method to use when putting the chemical into a refrigeration system. When using a refrigerant drum, vapor is released when the drum is in the upright position, and liquid is released when it is turned upside down.

Figure A-16 represents an undercharged system in which there is no leak. This could occur if the charge was incorrect from the factory or if the unit was serviced and an inadequate amount of refrigerant was put in. Both the high- and low-side pressures are lower than normal, and there is only a partial frost pattern on the evaporator. There would be a wide difference in the temperatures read at the inlet and at the outlet of the evaporator because, while there is some refrigerant in the system doing work as it enters the evaporator, there is not enough to chill the entire evaporator. Instead of a constant hissing as the refrigerant passes from the cap tube into the evaporator, the sound would come and go. The condenser would be slightly cooler than normal and the wattage would be low. Installing your gauges on this unit would show the low side operating in a vacuum rather than near 0 PSI and the high-side gauge would show lower

Figure A-13. A refrigerant recovery system. *(Courtesy Robinair).*

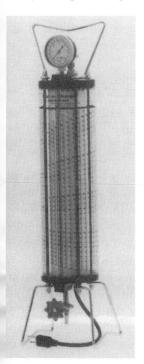

Figure A-14. A charging cylinder is used to correctly measure refrigerant into a system. *(Courtesy Robinair).*

Figure A-15. An electronic charging scale can be used to accurately charge a refrigeration system *(Courtesy Robinair).*

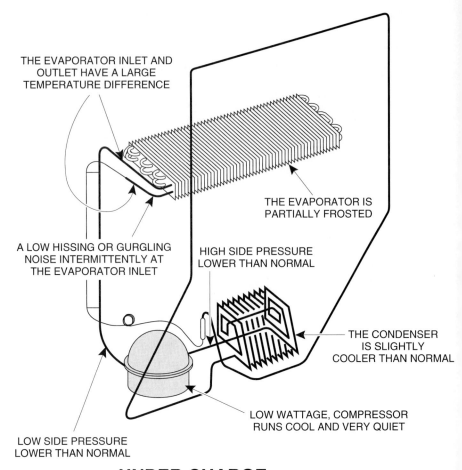

UNDER CHARGE
(NO LEAKS)

Figure A-16.   An undercharged system with no leak.

than the 126.6 PSI we calculated for our properly operating system in a 75° ambient.

One solution to this problem would be to install your gauges, recover the refrigerant, then measure in the correct charge.

Figure A-17 shows an undercharged system with a low-side leak. Many of the symptoms are the same as an undercharge without a leak, with the exception being the wattage draw and the high-side operating pressure. These factors are different because with a low-side leak, some air would be introduced into the system. Since air is a noncondensable, it affects the compressor as it does work and it causes the high-side pressure to rise.

To solve this problem, install your gauges and recover the refrigerant. Pressurize the system with dry nitrogen and use a soap solution to locate the leak.

Caution! *Never use oxygen to pressurize a system! Never open a high-pressure tank unless it is equipped with a pressure regulator. Failure to follow safety rules can result in an explosion that could cause serious injury or death.*

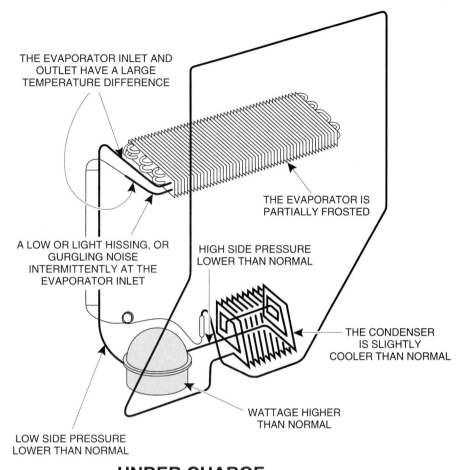

THE EVAPORATOR INLET AND
OUTLET HAVE A LARGE
TEMPERATURE DIFFERENCE

THE EVAPORATOR IS
PARTIALLY FROSTED

A LOW OR LIGHT HISSING, OR
GURGLING NOISE
INTERMITTENTLY AT THE
EVAPORATOR INLET

HIGH SIDE PRESSURE
LOWER THAN NORMAL

THE CONDENSER
IS SLIGHTLY
COOLER THAN NORMAL

WATTAGE HIGHER
THAN NORMAL

LOW SIDE PRESSURE
LOWER THAN NORMAL

UNDER CHARGE
LEAK (LOW SIDE)

Figure A-17. An undercharged system with a low-side leak.

If the leak cannot be found through this method, you can vent the nitrogen and use refrigerant to pressurize the system. You can then use an electronic leak detector to find the leak. Many electronic leak detectors are designed to sense a leak as small as $\frac{1}{2}$ ounce per year. One type of electronic leak detector is shown in Figure A-18.

Once the leak is found, recover the refrigerant. Repair the leak either by soldering the tube or joint, or by replacing the appropriate component (the evaporator or suction line); then install a new filter/drier, evacuate the system, and recharge with the correct amount of refrigerant.

Figure A-19 shows an illustration of a filter/drier and the proper method for installing the capillary tube. When preparing the filter/drier for installation, you can cut the condenser end with tubing cutters, but score the capillary tube end with a three-cornered file before snapping the end off the drier assembly, as shown in Figure A-20.

To evacuate a system, a vacuum pump such as the one shown in Figure A-21 must be connected. There are two schools of thought in using a vacuum pump and charging a system. Through one method,

Figure A-18. An electronic leak detector can be used to find small leaks in a refrigeration system. *(Courtesy Robinair).*

you would use the high- and low-side hoses on the appropriate access valves on the system, then connect the center hose to the vacuum pump, open both valves on the manifold gauges, and allow the vacuum pump to evacuate the system. After the proper vacuum is reached, you would close both sides of the gauges, disconnect the vacuum pump hose, and connect it to the refrigerant supply (charging cylinder or refrigerant drum on a scale).

At this point, it's important to understand that you have introduced air into the center hose you have disconnected and then reconnected. To remedy this, pressurize the center hose with refrigerant, then slowly unscrew the hose at the gauge assembly to purge the hose. After purging, you would proceed with charging the system by allowing liquid refrigerant to flow into the system by opening one side (according to most manufacturer's specifications, the high side) of the manifold gauge.

Another method recommended by some manufacturers is to use the three-hose gauge system with only one access. This allows you to connect the center hose to the refrigeration system, the high-side hose to the refrigerant supply, and the low-side hose to the vacuum pump. With this hookup, opening both gauges allows you to evacuate all hoses and the refrigeration system. Then, closing the low side only allows you to proceed with putting the refrigerant into the system through the high-pressure side of the gauge assembly. This method of using one access on the system is shown in Figure A-22.

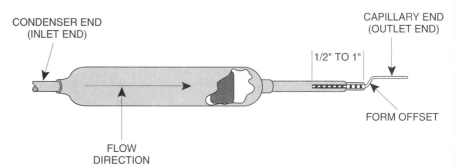

CONDENSER END
(INLET END)

CAPILLARY END
(OUTLET END)

1/2" TO 1"

FORM OFFSET

FLOW
DIRECTION

Figure A-19. The proper method of installing a capillary tube into a filter/drier.

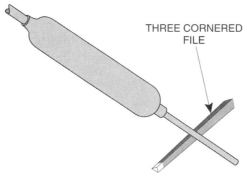

THREE CORNERED
FILE

Figure A-20. Using a three-cornered file to score the drier. This method is used in place of a tubing cutter on the capillary tube side of the drier.

Figure A-21. A vacuum pump is used to evacuate a refrigeration system. *(Courtesy Robinair).*

Figure A-23 shows an undercharged system with a high-side leak, which is similar to the undercharged system with no leak, with some minor differences. The noise level is even lower than the undercharged/no-leak system because refrigerant is being lost out of the high side of the system and no air is being drawn in since the compressor is maintaining a positive pressure. Eventually, a system operating with this problem would show no frost pattern at all on the

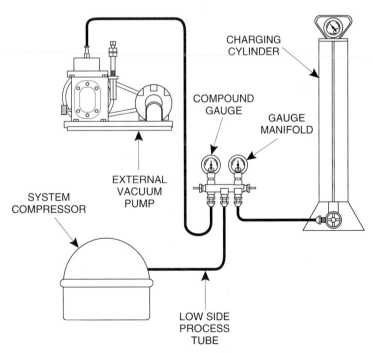

Figure A-22. One method of connecting a vacuum pump to a refrigeration system.

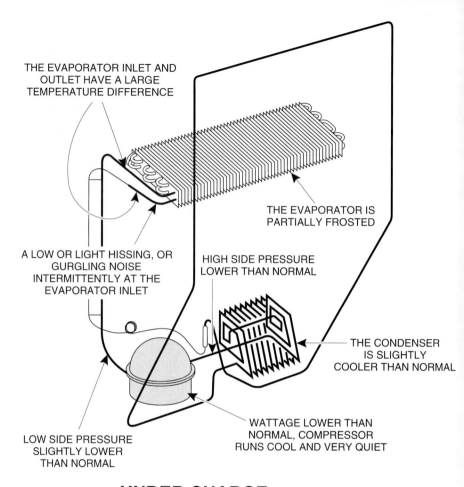

THE EVAPORATOR INLET AND
OUTLET HAVE A LARGE
TEMPERATURE DIFFERENCE

THE EVAPORATOR IS
PARTIALLY FROSTED

A LOW OR LIGHT HISSING, OR
GURGLING NOISE
INTERMITTENTLY AT THE
EVAPORATOR INLET

HIGH SIDE PRESSURE
LOWER THAN NORMAL

THE CONDENSER
IS SLIGHTLY
COOLER THAN NORMAL

WATTAGE LOWER THAN
NORMAL, COMPRESSOR
RUNS COOL AND VERY QUIET

LOW SIDE PRESSURE
SLIGHTLY LOWER
THAN NORMAL

UNDER CHARGE
LEAK (HIGH SIDE)

Figure A-23. An undercharged system with a high-side leak.

evaporator, and the noise level at the inlet of the evaporator would be nil.

If there is any refrigerant left in this system, recover it. Locate and repair the leak in the manner previously described, install a new filter/drier, evacuate, and recharge.

Figure A-24 represents a system with a partial restriction on the high side. The restriction could be in the condenser or yoder loop due to a kinked or pinched tube, but more than likely would be in the filter/drier. The noise at the evaporator inlet would indicate constant refrigerant flow, but the volume would be too low to sustain a properly chilled evaporator. The high-side pressure would be higher than normal due to the fact that refrigerant would not be flowing freely to the evaporator. Also, the wattage would be lower than normal because of the low volume of refrigerant in the suction line, preventing the compressor from having enough work to do.

One method to follow in solving this problem would be to install your gauges and recover the refrigerant, making sure you are recovering from both the high- and low-side access valves. If you suspect

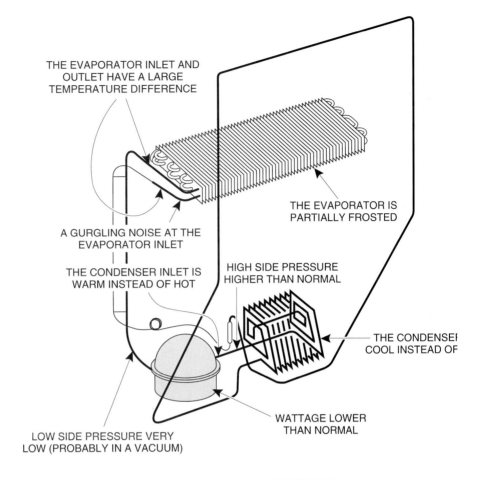

THE EVAPORATOR INLET AND
OUTLET HAVE A LARGE
TEMPERATURE DIFFERENCE

THE EVAPORATOR IS
PARTIALLY FROSTED

A GURGLING NOISE AT THE
EVAPORATOR INLET

HIGH SIDE PRESSURE
HIGHER THAN NORMAL

THE CONDENSER INLET IS
WARM INSTEAD OF HOT

THE CONDENSE[
COOL INSTEAD OF

WATTAGE LOWER
THAN NORMAL

LOW SIDE PRESSURE VERY
LOW (PROBABLY IN A VACUUM)

A PARTIAL RESTRICTION
(HIGH SIDE)

Figure A-24. A refrigeration system with a partial restriction on the high-pressure side.

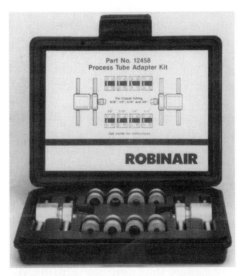

Figure A-25. A process tube adapter kit is used for temporary access to a refrigeration system. *(Courtesy Robinair).*

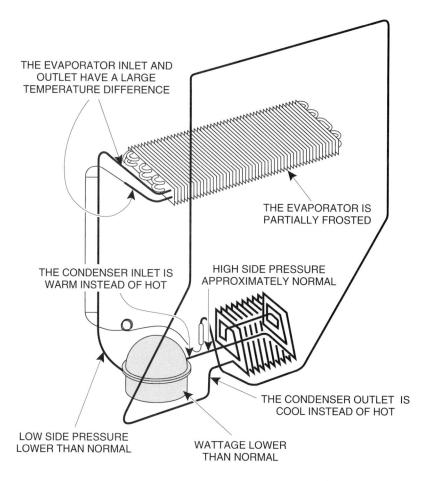

THE EVAPORATOR INLET AND
OUTLET HAVE A LARGE
TEMPERATURE DIFFERENCE

THE EVAPORATOR IS
PARTIALLY FROSTED

THE CONDENSER INLET IS
WARM INSTEAD OF HOT

HIGH SIDE PRESSURE
APPROXIMATELY NORMAL

THE CONDENSER OUTLET IS
COOL INSTEAD OF HOT

LOW SIDE PRESSURE
LOWER THAN NORMAL

WATTAGE LOWER
THAN NORMAL

A PARTIAL RESTRICTION
(LOW SIDE)

Figure A-26. A refrigeration system with a partial restriction on the low-pressure side.

the restriction is in a section of the condenser, isolate this section of the system by cutting the connecting tubing and installing a process tube adapter tool that slips over the end of the open tube, allowing for the installation of a charging hose. A process tube adapter kit is shown in Figure A-25.

Use dry nitrogen to blow through the component and make a judgment as to whether it is clear. Repeat the procedure on any section of the system you suspect as restricted. One common area of a partial high-side restriction is the filter/drier itself. Replacing it in the manner previously described, evacuating the system, and recharging would solve the problem.

In Figure A-26, you see a system with a partial restriction on the low-pressure side. With the condition, the low-pressure reading would be in a vacuum, and the high-pressure reading would be slightly higher than normal. Putting your hand on the discharge line, you would find the temperature to be lower than that of a normally operating unit. Wattage would also be lower than normal due to the fact

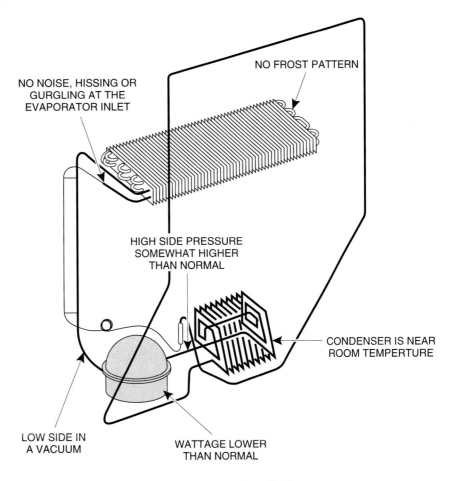

NO NOISE, HISSING OR GURGLING AT THE EVAPORATOR INLET

NO FROST PATTERN

HIGH SIDE PRESSURE SOMEWHAT HIGHER THAN NORMAL

CONDENSER IS NEAR ROOM TEMPERTURE

LOW SIDE IN A VACUUM

WATTAGE LOWER THAN NORMAL

A COMPLETE RESTRICTION

Figure A-27. A refrigeration system with a complete restriction (solid contaminant).

that the volume of refrigerant in the suction line is low and the compressor is not doing the maximum amount of work.

Solving this problem would require following the procedures already described in regard to isolating and pressurizing suspected areas of restriction. To complete the repair, replace the filter/drier, evacuate the system, and recharge.

Figure A-27 shows a complete restriction. Since the restriction is total, the low-side reading would be in a deep vacuum, probably near 25 inches on your gauge, and the high-side pressure would be higher than normal since the compressor is pumping up against a restriction. This condition could occur in the filter/drier. If this is the case, follow the procedures previously described to complete the repair.

In Figure A-28, a system suffers from a restriction due to moisture in the system. This system would operate normally for a period of time; then the moisture would freeze at the evaporator inlet, restricting the refrigerant flow. After a period of no cooling, the ice would thaw, allowing the refrigerant to flow again until freezing again caused the restriction.

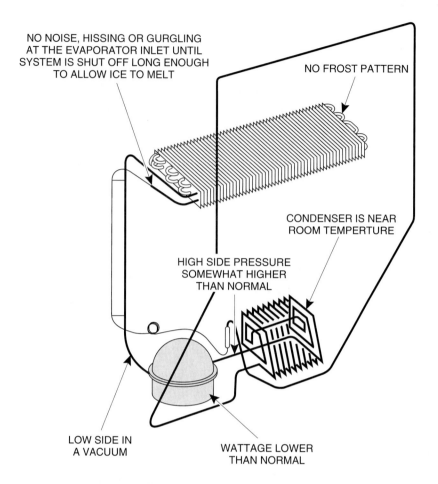

NO NOISE, HISSING OR GURGLING
AT THE EVAPORATOR INLET UNTIL
SYSTEM IS SHUT OFF LONG ENOUGH
TO ALLOW ICE TO MELT

NO FROST PATTERN

CONDENSER IS NEAR
ROOM TEMPERTURE

HIGH SIDE PRESSURE
SOMEWHAT HIGHER
THAN NORMAL

LOW SIDE IN
A VACUUM

WATTAGE LOWER
THAN NORMAL

A COMPLETE RESTRICTION
(DUE TO MOISTURE AT END
OF CAPILLARY TUBE)

Figure A-28.   A refrigeration system with a restriction caused by moisture.

Installing your gauges on this unit would yield the conditions shown unless, of course, you happened to be testing the system at a time just after the thawing of the restriction. If moisture is the problem in a system, it will most likely freeze again in approximately 10 minutes.

To effect this repair, install a new filter/drier and proceed with evacuation. If there is moisture in a system, pulling a vacuum near the maximum possible of 29.94 inches will cause the moisture to boil into a vapor and be drawn out of the system via the vacuum pump. Water boils at 212° Fahrenheit at atmospheric pressure of 14.7 PSI. As you drop the pressure, the boiling point drops also. At 27 inches of vacuum, the boiling point of water is lowered to near 75° Fahrenheit. What this means is that you should make sure the refrigeration system components are warmed up and the vacuum is pulled as low as possible to draw the moisture out of the system.

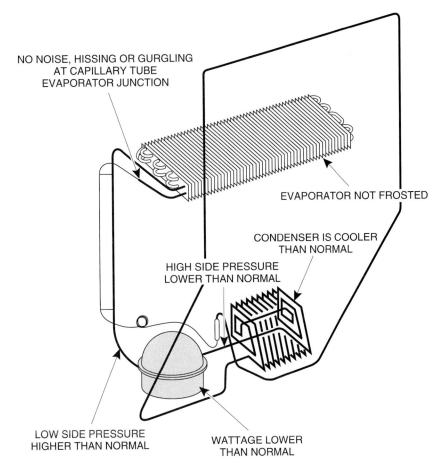

INEFFICENT COMPRESSOR

Figure A-29.   A refrigeration system with an inefficient compressor.

To complete the repair, dial in or measure in the correct amount of refrigerant; then monitor the operation of the unit for a reasonable length of time to ensure that it is operating properly.

Figure A-29 represents a system with an inefficient compressor. In the case of a compressor valve failure, the low-pressure reading would not be near 0 PSI as it would on a normally operating system. Instead, it would be higher than normal. Just how much higher it would be depends on the extent of the valve failure. It's common to find the low side running at 15 to 20 PSI in case of a valve failure.

The high-side operating pressure would be lower than the 126.6 PSI described on our normally operating system. As with the low side, the extent of the failure would determine just how much lower than normal the operating pressure would be. In some cases of a total valve failure, the pressure differential between the low side and the high side can be very close.

To effect this repair, recover the refrigerant, cut the tubing connected to the failed compressor, and install a new compressor and filter/drier. Evacuate the system and measure in the correct refrigerant charge.

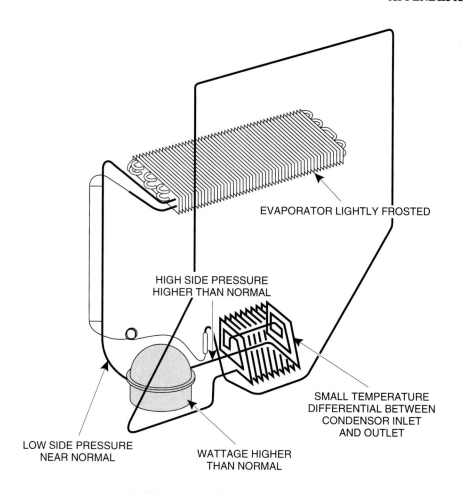

EVAPORATOR LIGHTLY FROSTED

HIGH SIDE PRESSURE
HIGHER THAN NORMAL

SMALL TEMPERATURE
DIFFERENTIAL BETWEEN
CONDENSOR INLET
AND OUTLET

LOW SIDE PRESSURE
NEAR NORMAL

WATTAGE HIGHER
THAN NORMAL

AIR IN SYSTEM

Figure A-30. A refrigeration system containing noncondensibles.

In Figure A-30, air is in the refrigeration system. This condition usually exists when a system has been improperly serviced in some way, such as not purging a hose before adding refrigerant. The high-side pressure would be higher than normal due to the fact that air is a noncondensible. This fact would also cause the wattage to be higher than normal. The frost pattern on the evaporator would be complete but not as normal due the fact that the air in the system would be preventing proper heat transfer into the refrigerant.

To complete this repair, recover the refrigerant, install a new filter/drier, evacuate, and recharge.

A.5 R-12 AND HFC-134A

Since the late 1930s, R-12 has been the refrigerant of choice for use in refrigerators and freezers. With the EPA Clean Air Act of 1990, a major change in the laws surrounding the use of chloro-fluorocarbon (CFC) chemicals as refrigerants and their availability for purchase by technicians has undergone a great deal of change. R-12 has been targeted for production phase-out due to fact that it is listed as causing the greatest damage to the ozone layer, both because of its chemical makeup and the volume of its use.

An alternate refrigerant, HFC-134A, has been developed as a replacement for R-12. While it is similar in chemical and physical characteristics, it is not considered to be a drop-in replacement for R-12. Manufacturers are not recommending replacing R-12 with HFC-134A. One of the problems encountered is the compatibility of the new refrigerant with the oils commonly used in R-12 systems. Research showed that HFC-134A would not work well with the traditional mineral or synthetic oils used in refrigerator compressors.

In addition, the boiling point of HFC-134A is slightly higher than R-12. R-12 boils at –21.6° Fahrenheit and HFC-134A boils at –15.7° Fahrenheit. This means that HFC-134A will have a generally higher compression ratio than R-12, and, to achieve the same cooling capacity, a larger compressor would be required. Since oil compatibility and compressor capacity are two of the critical factors to be considered, changing a R-12 refrigerator to HFC-134A would not be practical.

One manufacturer, DuPont, has developed a line of refrigerants as a replacement for CFCs. Under their registered trademark, CFCs were identified as Freon, and the new HFC refrigerants are identified

Temp. (°F)	Vapor Pressure*			
	SUVA™ Cenri-LP HCFC 123	SUVA™ Chill-LP HCFC 124	SUVA™ Freez-HP HFC 125	SUVA™ Cold-MP HFC 134a
−100	29.9	29.2	24.4	27.8
−90	29.8	28.8	21.7	26.9
−80	29.7	28.2	18.1	25.6
−70	29.6	27.4	13.3	23.8
−60	29.5	26.3	7.1	21.5
−50	29.2	24.8	0.3	18.5
−40	28.9	22.8	4.9	14.7
−30	28.5	20.2	10.6	9.8
−20	27.8	16.9	17.4	3.8
−10	27.0	12.7	25.6	1.8
0	26.0	7.6	35.1	6.3
10	24.7	1.4	46.3	11.6
20	23.0	3.0	59.2	18.0
30	20.8	7.5	74.1	25.6
40	18.2	12.7	91.2	34.5
50	15.0	18.8	110.6	44.9
60	11.2	25.9	132.8	56.9
70	6.6	34.1	157.8	70.7
80	1.1	43.5	186.0	86.4
90	2.6	54.1	217.5	104.2
100	6.3	66.2	252.7	124.3
110	10.5	79.7	291.6	146.8
120	15.4	94.9	334.3	171.9
130	21.0	111.7	380.3	199.8
140	27.3	130.4	430.2	230.5
150	34.5	151.0	482.1	264.4
160	42.5	173.6		301.5
170	51.5	198.4		342.0
180	61.4	225.6		385.9
190	72.5	255.1		433.6
200	84.7	287.3		485.0
210	98.1	322.1		540.3
220	112.8	359.9		
230	128.9	400.6		
240	146.3	444.5		
250	165.3	491.8		
260	185.8			
270	207.9			
280	231.8			
290	257.5			
300	285.0			

*Vapor pressures are shown as PSIG.

Figure A-31. A temperature/pressure chart for refrigerant HFC-134a.

under the brand name SUVA. Figure A-31 shows a temperature/pressure chart for the replacement refrigerants.

A.6 SWAGING AND FITTING TUBING

When replacing compressors, installing filter/driers, and performing other repairs on refrigeration systems, you will have to fit tubing together. *Swaging* ensures a good fit. With swaging tools and a block to hold the tubing, you can fit two tubes of the same size together to make a joint. An example of a swaging tool is shown in Figure A-32, and a swaging block is shown in Figure A-33. The swaging block is sized to accept various sizes of tubing and hold them tightly while you drive the swaging tool down into the tubing opening with a hammer.

A.7 SOLDERING

Caution! *Always wear proper and approved safety glasses when working with a torch. Have an approved dry-type fire extinguisher on hand when using any type of gas-operated torch.*

All refrigeration system joints must have a proper fit. Manufacturers recommend that the clearance between the tubes be from 0.001 to 0.006 inch. You won't be able to actually measure this, but keep in mind that, from a common sense point of view, you don't

Figure A-32. A swaging tool is used to accomplish the task of fitting tubing together. *(Courtesy Robinair).*

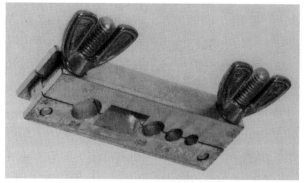

Figure A-33. A combination swaging and flaring block. *(Courtesy Robinair).*

Figure A-34. An acetylene torch is used for proper soldering of refrigerant system tubing connections. *(Courtesy Robinair).*

want two tubes that fit too closely together, which would prevent solder from flowing down into the joint where it has to go. Nor do you want two tubes that fit together too loosely, resulting in a sloppy joint. Tubing joints should overlap about the distance of their diameter.

Before fitting two tubes together, remove dirt, grease, or oxidation with a fine steel wool or, better yet, an appropriate abrasive cloth. After cleaning the tubing, apply a thin coat of flux that is compatible with silver solder.

To effectively solder joints, you need to use a torch of adequate capacity so that the copper tubing can be quickly heated to the required temperature. Silver solder, which has a melting point near 1100° Fahrenheit, is recommended for use on refrigerator and freezer system joints. Some technicians prefer to use an oxy/acetylene torch assembly while others use an acetylene-only system, such as the one shown in Figure A-34.

When fitting tubing together, be sure to align the tubing so that there is no stress on the joint. It's also a good idea to push two tubes coming together from opposite directions past each other before actually fitting them together. This will keep them together under a small amount of pressure and prevent them from slipping apart when you apply heat from the torch.

When making a good silver solder joint, if it is properly cleaned and fluxed, the solder will flow readily into the joint. You don't need to use a lot of solder on a joint, just enough to make a good bond.

It's also a good idea to carry a piece of heat-resistive material to place on the floor when you're using a torch in a customer's home. This will prevent damage to the floor. In some cases, you may also want to use a wet cloth to prevent heat from conducting to areas

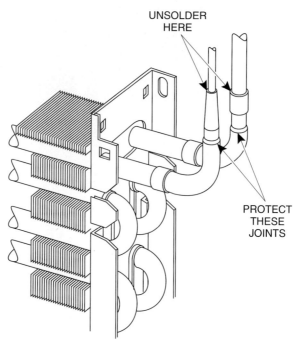

UNSOLDER
HERE

PROTECT
THESE
JOINTS

Figure A-35. When working near aluminum components, care must be taken to protect the components and joints that would be damaged by high temperatures.

other than the soldered joint. If you were replacing an evaporator, for example, you would need to protect the aluminum component from excessive heat while soldering the copper stubs to the capillary tube and suction line. Figure A-35 shows a finned-type evaporator and the joints that must be protected from excessive heat while unsoldering and resoldering.

A.8 REFRIGERANT RECOVERY EQUIPMENT

To be in compliance with EPA rulings governing the venting of refrigerants into the atmosphere, you'll have to work with refrigerant recovery equipment. Some manufacturers recommend the use of a refrigerant recovery bag, which is fundamentally a heavy duty plastic bag into which the technician allows the refrigerant to vent naturally prior to repairing the refrigeration system. There are also other natural-flow refrigerant recovery units designed for use on systems, such as refrigerators and freezers that use what is considered to be a relatively small refrigerant charge. The most common type of recovery equipment, though, is the mechanical system. Basically a refrigeration system by nature, it contains a compressor to pull the refrigerant from a system and uses a condenser to complete the transfer of refrigerant in a liquid form. A mechanical refrigerant recovery system is shown in Figure A-36.

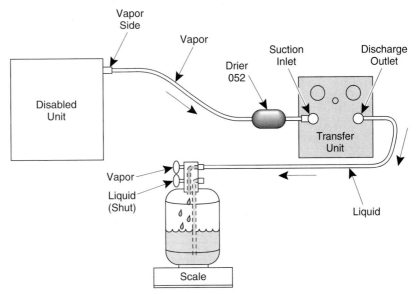

Figure A-36. A refrigerant recovery system draws refrigerant from the disabled unit and transfers it to a storage tank.

A.9 BUILDING YOUR OWN REFRIGERANT RECOVERY SYSTEM

It is possible to build rather than buy a refrigerant recovery system. But putting one together that is effective and safe to operate takes some effort and working with refrigeration components you would not normally use in a domestic refrigeration system. In addition to the fan-cooled condensing unit consisting of the compressor and condenser, a viable unit would contain hand valves that would allow for the connection of charging hoses, an accumulator/oil trap on the suction side of the system, a crankcase pressure regulator, sight glasses, replacable filter/driers, and a high-pressure cut-out switch for safety. The compressor on a shop-built refrigerant recovery system would also have to be modified by drilling a hole in its bottom and adding an oil drain. This would allow for the changing of the compressor oil, something you would have to do on a regular basis to keep your recovery system functioning properly.

Figure A-37 shows an illustration of the basic components of a shop-built refrigerant recovery system. This illustration is part of a set of complete plans offered by Set Point Heating and Air Conditioning of Granada Hills, California. Referred to as the Cheap One because the estimated construction cost is $350, it illustrates the direction of refrigerant flow in a recovery system.

To use the system, you would connect the manifold gauges and charging hoses from the disabled unit to the suction valve shrader and draw the refrigerant from the refrigerator. The direction of refrigerant flow is shown through the suction accumulator, suction line filter/drier, suction line sight glass, and on through the crankcase pressure regulator to the compressor.

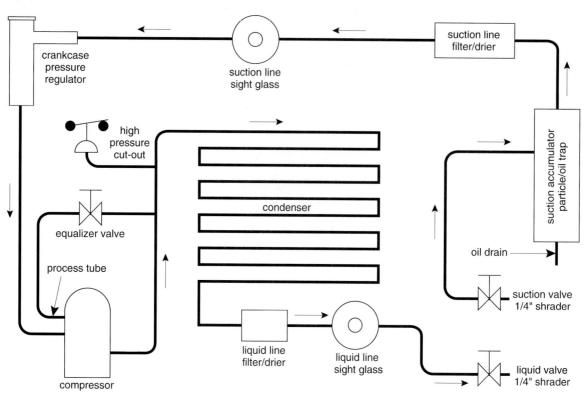

Figure A-37. An illustration of a shop-built refrigeration system known as the Cheap One.

As in a standard refrigerant system, the compressor raises the pressure of the vapor, and the condenser allows the change in state from a vapor to a liquid. In its liquid form, the refrigerant passes through the liquid line filter/drier and liquid line sight glass before being allowed to leave the recovery system through the liquid valve shrader. Connecting a hose from this end of the system to a storage tank would allow you to capture the refrigerant in liquid form.

Caution! *Always exercise extreme care when using any refrigerant recovery system and transferring into a storage drum. Overfilling will cause an explosion that can result in serious injury or death! Never use a recovery container that is not designed and approved for the purpose of recovering refrigerant. The construction and operation of a refrigerant recovery system require detailed instruction and proper training.*

Refrigerator/Freezer Electrical Components

B.1 TEMPERATURE CONTROL

To maintain design temperature, all refrigerators and freezers must have a method of initiating and terminating the refrigeration system. The answer is a control that senses temperature, and then either makes or breaks the circuit to the compressor and other refrigeration components. Your customer may refer to it as the thermostat, some technicians refer to it as the cold control, but most manufacturers prefer to use the term *temperature control*.

The control itself is nothing more than a set of contacts that are controlled by a sensing bulb charged with refrigerant. The sensing bulb (sometimes referred to as a *feeler tube* or *capillary tube*) may be fastened to the evaporator, as in the case of a conventional or cycle defrost unit. Or it may be an air-sensing control, which is the common method of operation in a frost-free refrigerator. From the customer's point of view, the control consists of a knob with numbers. From the technician's point of view, it's a component with two wiring connections, one for power into the switch and the other to allow for power out to the cooling system components. A cutaway view of a temperature control is shown in Figure B-1.

The sensing tube is charged with refrigerant. You'll recall that there is a temperature/pressure relationship between refrigerants, meaning that, if the temperature rises, the pressure rises and, when the temperature drops, the pressure drops. This fundamental idea applies to the temperature control. When the temperature of the evaporator or the interior cabinet air temperature rises, the pressure in the sensing tube rises, overcoming the spring pressure inside the control body and allowing the contacts inside the control to make.

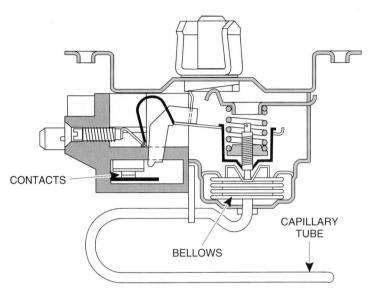

Figure B-1. A temperature control is a switch that is activated by pressure in its sensing tube.

When the temperature drops due to the work being done by the cooling system components, the pressure in the feeler tube drops, allowing the contacts inside the control body to break the circuit to the cooling system components.

A temperature control can fail in one of two ways: Either it will fail in the open position and fail to complete the circuit to the cooling components, or it will fail in the closed position and fail to break the circuit to the cooling components, even though the desired temperature has been reached. Failing in the open position, due either to loss of charge in the sensing tube or to corroded contact points inside the control, is the more common kind of failure. This failure results in a customer complaint of "not cooling, not making any noise."

When responding to this type of complaint, diagnosing the problem and testing the control can be accomplished as follows:

Step 1: Unplug the service cord from the receptacle. Failing to do this could result in electric shock or tripping a circuit breaker when you allow a hot wire to come in contact with the inner liner of the refrigerator if it happens to be of porcelain on steel construction.

Step 2: Remove the knob from the control. Most simply pull straight off; some older models may require the removal of a screw through the center of the knob.

Step 3: Remove the mounting screws that secure the control to the inner liner, and pull the control out through the access opening or remove the housing that surrounds the control.

Caution! *Careful if you're working with a model that requires removal through an opening in the inner liner. The opening may be small, and it may require some twisting and turning to get the control out through the opening. Take care not to damage the control or the liner.*

Step 4: Disconnect one wire from the control, set your ohmmeter on the R×1 scale, and test for continuity between the two terminal connections of the control. If the control is turned up and calling for cooling, and the refrigerator is warm, the meter should indicate continuity. If the needle of the meter doesn't move or if the digital readout doesn't indicate continuity, the control has failed. If continuity is indicated, the control, in 99.9 percent of the cases is okay. (Rarely, your sensitive meter will indicate continuity, but when the control is put to actual use, the contact points inside can't stand up to the current flow.)

To further test a cold control and refrigerator/freezer electrical system, follow Steps 1 through 3, then proceed as follows:

Alternate Step 4: After gaining access to the control electrical connections, install a jumper between the two wires. Make sure the wires are not touching metal, and plug the service cord back into the receptacle. If the unit wouldn't start before, but does with the jumper installed, the control is bad and must be replaced.

Another hot test you can perform in this situation is to use your voltmeter across the terminals of the control (with the wires still connected). If the control is turned up and calling for cooling and the refrigerator is warm, a good control will show no voltage across the terminals. If the control has failed, 120 volts will be read across the terminals.

Caution! *Always use extreme care when working with hot circuits.*

Temperature controls have cut-in and cut-out temperatures that varies widely depending on whether the unit is a conventional refrigerator in which the control sensing tube is fastened to the freezer that makes up the evaporator, fastened to the cooling plate section of the evaporator in a cycle defrost unit, or in the air stream from the freezer section of a frost-free unit. The control is designed specifically for the model in which it is used, and the replacement control must be a match for the one being replaced.

The control will have, depending on the manufacturer, one or two adjustment screws for the cut-in and cut-out temperatures, but these are *only* to be used when adjusting the control for a change in altitude. Figure B-2 shows a Cutler-Hammer type of control with two adjusting screws. A GE control also has a cut-in/cut-out adjustment, but it is accomplished with one screw.

Cut-in cut-out adjustment is necessary when the refrigerator is installed at a high altitude, which affects the evaporating temperature of the refrigerant in the sealed system. To compensate for this situation, the adjustment screws on the temperature control are set according to specific manufacturer's instructions for the particular model you're servicing. *Never* attempt to adjust the cut-in/cut-out

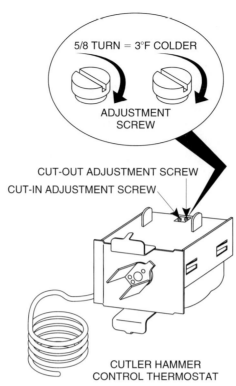

Figure B-2. A temperature control cut-in/cut-out temperature must be adjusted for altitude according to manufacturer's instructions.

screws in an effort to solve a customer complaint of improper operating temperatures. In the end, you'll do more harm than good.

B.2   CURRENT RELAY AND PTC START DEVICE

In refrigerators and freezers, a current relay or a solid state device known as a *PTC* (*positive temperature coefficient*) *relay*, sometimes referred to as a *PTCR*. In either case, it performs the same function in momentarily allowing the start windings of the compressor motor to be energized until the motor reaches approximately 75 percent of its running speed. Since the coil of the current relay is wired in series with the run winding of the compressor, the current draw through this winding governs whether or not the contacts inside the relay are in the open or closed position. For this reason it's referred to as a *current relay*. You may find some manufacturers referring to it as an *amperage relay*.

Most current relays plug onto the pins that serve as electrical connections to the motor windings of the compressor. One common method of wiring is the hot wire connected to the relay and the relay plugged onto the start and run winding pins of the compressor. A complete circuit through the compressor windings is accomplished by the neutral wires being connected to the common pin on

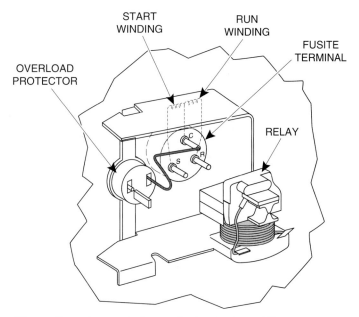

Figure B-3.   A current relay, overload protector, and the compressor electrical terminals.

the compressor. Remember that not all manufacturers wire their units in this fashion. Refer to Chapter Seven for a review of a simplified wiring diagram that shows the compressor windings. Figure B-3 shows the most common configuration of the pins on a refrigerator compressor.

To remove the current relay, proceed as follows:

Step 1: Unplug the power cord of the unit.

Step 2: Remove the relay cover clip and relay cover.

Step 3: Pull straight back to remove the relay from the compressor pins.

A current relay can fail in several different ways, ranging from an open coil to contacts that won't close to contacts that won't break. In either case, failure of the current relay would result in a

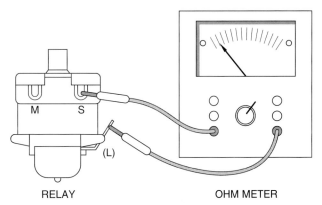

Figure B-4.   Testing a current relay in the upright position.

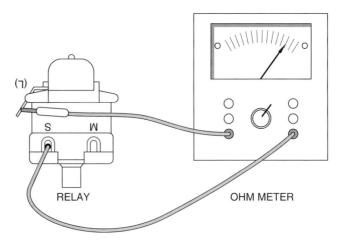

Figure B-5. Testing a current relay in an inverted position.

customer complaint of "not cooling." Some observant customers may be able to tell you that the refrigerator doesn't sound as though it's running and that it clicks every few minutes.

Once you remove the relay, you can use an ohmmeter to test it for proper performance. Holding the relay in its upright position and with your meter set on R×1 scale, touch the leads of the meter to the L (line) and S (start) terminals of the relay. There should be no continuity reading between these two terminals with the relay in the upright position. Figure B-4 shows a current relay continuity test.

You can simulate the relay operating and closing the contact points inside by turning the relay upside down. As shown in Figure B-5, when the relay is in this position, the meter should show continuity. If no continuity is shown, the relay must be replaced.

The PTC start relay is not an electromechanical device like the current relay. The PTC start relay is a solid state device that changes resistance as the temperature changes. In this case, positive temperature coefficient means that, as the temperature goes up, the resistance goes up. When the temperature goes down, the resistance goes down. At the instant of start, the PTC device has very low resistance and allows a circuit to the start winding. The current draw of the

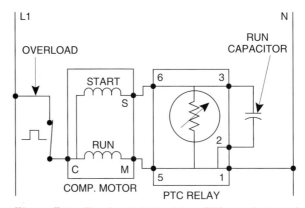

Figure B-6. The electrical circuit for a PTC start device and run capacitor.

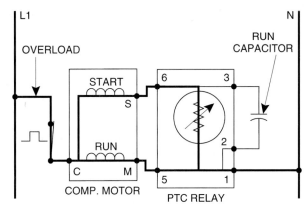

Figure B-7. The PTC circuit at the instant of start. The bold lines show the power being applied to both windings.

start winding causes the temperature of the start device to rise, and as a result the resistance rises, ultimately breaking the circuit to the start winding. A PTC device is also commonly used in conjunction with a run capacitor, as shown in the electrical circuit diagram in Figure B-6.

As you can see from the diagram, L1, or the hot side of the line, is wired through an overload protector and connected to common on the compressor. The N side of the line is wired to terminal #1 on the relay. This power leg will energize both the run and start winding due to the low resistance of the PTC device at the instant of start. The symbol for the variable resistor is located in the middle of the PTC relay assembly.

At the instant of start, as shown in Figure B-7 but the bold lines on the electrical circuit, N now delivers voltage through terminal #5 to M (the run winding on the compressor) and through terminal #6 to S (the start winding). You'll note that at this time there is no power through the run capacitor.

Figure B-8 shows the circuit after the resistor has reacted to the temperature rise and the resistance has increased. Now the circuit

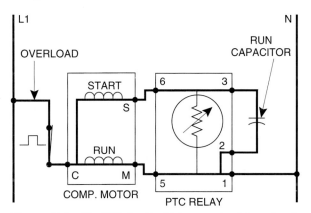

Figure B-8. The PTC circuit with increased resistance breaking the circuit to the start winding. The bold lines show the circuit to the run winding and through the run capacitor to the start winding.

directly to the start winding has been broken (while it is maintained through the run winding), and instead the circuit to the start winding runs through the run capacitor. The run capacitor is wired in series with the start winding to allow the compressor to run more efficiently.

Exact operating characteristics vary from manufacturer to manufacturer, but the PTC device used in our example has a resistance of 3 to 12 ohms when at room temperature. At higher operating temperatures, the resistance increases very quickly to 10 to 20 K ohms.

To test a PTC relay, proceed as follows:

Step 1: Unplug the refrigerator from the receptacle.

Step 2: Remove the relay cover retaining clip, and remove the plastic relay cover.

Step 3: Disconnect the wires connected to the relay, and pull the relay off the compressor terminals.

Step 4: Check with an ohmmeter set on the R×1 scale between the appropriate terminals (in our example, terminals 2 and 3). The resistance should be within the manufacturer's design parameters, usually between 3 and 12 ohms at room temperature. A shorted relay with read 0 resistance. If the relay is open, the meter will show no reading on the meter.

B.3 OVERLOAD PROTECTOR

An *overload protector* is a bimetal device wired in series with one side of the line to the compressor. In the event that the compressor overheats due to improper voltage, failure of the condenser fan motor or loss of air flow over a fan-cooled condenser to a dirt build-up, the overload protector will react by opening and breaking the

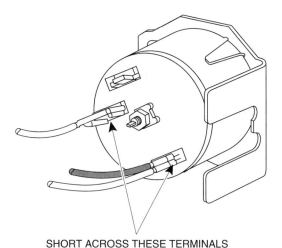

SHORT ACROSS THESE TERMINALS

Figure B-9. The overload protector used in conjunction with the current relay or PTC start device.

circuit to the compressor. This device, sometimes referred to as a *guardette*, is held in position near the compressor by a spring clip.

An overload protector may fail in the open position which would result in a complaint of "no cooling." One method of testing a suspected open overload protector would be to jump across the terminals of the device, effectively eliminating it from the circuit. Figure B-9 shows one type of overload protector and the method for jumping it out of the circuit.

B.4 COMPRESSOR MOTOR WINDINGS

The compressor motor contains two windings, the run winding and the start winding. The most common configuration of the windings electrical connections is shown in Figure B-10. Common is on the top, run on the right, and start on the left.

To test the motor windings of a compressor, proceed as follows:

Step 1: Unplug the refrigerator from the wall outlet, remove the relay cover, relay, and overload assembly.

Step 2: Set your ohmmeter on the R×1 scale, and test between the start and run terminals. This test should yield your highest resistance reading.

Step 3: Test between the common and start terminals. This reading will be lower than the start to run test but higher than the common to run test.

The actual readings you'll get will vary from one model of refrigerator to another, but it is usual to read between 1 and 1.5 ohms from common to run and between 8 and 15 ohms from common to start. Reading 1 ohm on the run winding and 8 ohms on the start winding would give you a reading of 9 ohms from start to run.

A compressor winding can fail "open," which would yield no reading on the meter, or compressor windings can be "shorted,"

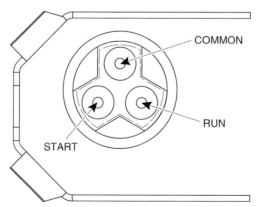

Figure B-10.   The most common configuration for compressor terminals common, run, and start.

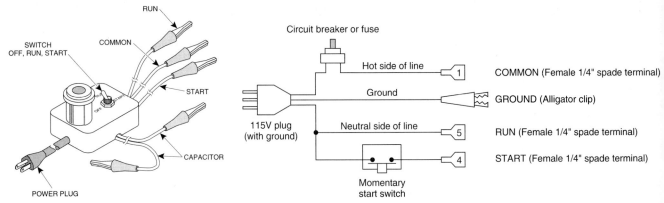

Figure B-11.   A compressor test kit.

which would yield the same resistance readings between all three terminals. A compressor can fail with a "grounded" condition, which would yield a meter reading from any electrical terminal to cabinet ground. An open, shorted, or grounded reading requires that the compressor be replaced.

B.5   COMPRESSOR TEST CORD

When troubleshooting a compressor and related starting components, one method of testing the compressor is to use a compressor start kit. Test sets, such as the one shown in Figure B-11, are available for purchase, or you may choose to build your own.

Use of the test set allows you to eliminate the relay and overload protector from the circuit and start the compressor directly from a power source. It also allows you to use an ammeter to test for proper current draw of the motor windings. If the compressor cannot start with the power applied directly from the test set, the compressor is stuck or has failed electrically. It must be replaced.

When building a compressor test set, it should be properly fuse protected. L1 should be wired through the fuse directly to the common alligator clip. L2 should be wired directly to the run alligator clip and should also be wired through a momentary switch to the start alligator clip. For ease of operation, an on/off switch should be included.

B.6   DEFROST TIMER

The *defrost timer* is a motor-driven device used to periodically energize the defrost heater and clear the finned type of evaporator used in a frost-free refrigerator of frost build-up. Without this procedure, the operating efficiency of the refrigerator would be affected because the air being circulated in a frost-free refrigerator must pass in close contact with the evaporator tubing for proper heat absorption from the air.

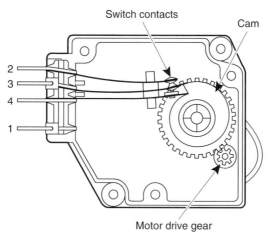

Figure B-12. A defrost timer uses a cam and gear assembly to initiate and terminate the defrost and run modes.

One method of wiring the defrost timer motor is for constant run time, while the other method is to wire it for cumulative run time. Wiring with the *cumulative* method means that the timer motor doesn't advance the timer toward the defrost mode unless the temperature control is calling for cooling. On a *constant run time* system, a refrigerator would defrost every 8 hours (or whatever the design of that particular unit), no matter how much or how little the refrigerator runs. On the cumulative run time system, the timer would reach the defrost mode only after 8 hours of compressor run time. The cumulative run time method of wiring a timer is the more common.

Figure B-12 shows a common method of construction of a defrost timer. The timer can be located in several different places on the refrigerator. You may find one inside the fresh food compartment behind a light guard, while another will be mounted on a bracket behind the toe plate of the unit.

Through a motor and cam assembly, a set of contacts provides a circuit to the cooling system components in the refrigerator run mode, and another set of contacts provides a circuit to the defrost heater in the defrost mode. When the defrost mode is energized, the cooling mode is deenergized. Refer to Chapter Seven for a review of wiring methods for a defrost timer, run cycle components, and defrost cycle components.

You'll run into some variables when working with different makes and models of refrigerators. A sample of some of them is shown in Figure B-13.

Defrost timers can fail in a variety of ways and cause a variety of symptoms and conditions. A timer motor that stalls in the run mode will allow a frost build-up to occur on the evaporator, which will eventually create higher-than-normal temperatures in the freezer and fresh food sections (usually in about 3 days). The customer's

The following diagrams indicate how the various refrigeration manufacturers mark the terminal boards of their timers. Switches are shown in the refrigeration position.

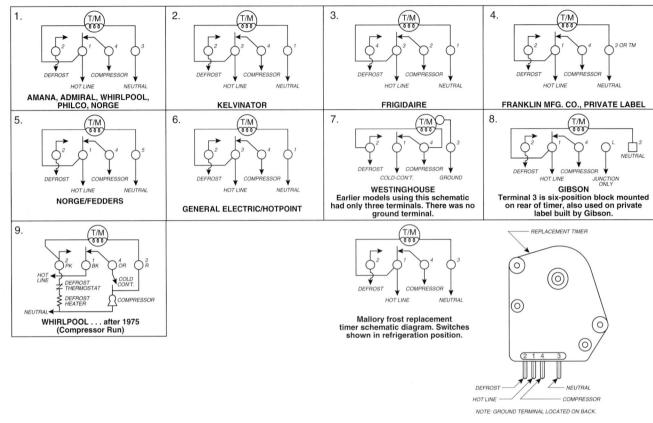

Figure B-13. Different manufacturers use different methods of wiring defrost timers.

complaint in this case would be "running but not cooling." To test for this condition of stuck-in-the-run mode, proceed as follows:

Step 1: Locate the timer and advance it manually in a clockwise direction until the compressor shuts off.

Step 2: Use either an ammeter around the appropriate wire in the wire harness or a volt/wattmeter to test for current draw in the defrost mode. If the current draw in the defrost mode is within the specifications of that unit (usually in the neighborhood of 3 to 5 amps) and the unit successfully defrosts, the defrost timer must be replaced. This diagnosis is accomplished by your doing manually what the timer should have been doing automatically.

Another possible failure of a defrost timer is the motor stalling in the defrost mode. This condition would result in the customer's complaint being "not cooling, not making any noise." To test for this condition, locate the timer and advance it clockwise. This advance would be for a very short distance since the defrost mode is much shorter than the run mode. If the unit starts and runs when you make this test, the timer must be replaced because it was not advancing automatically into the run mode.

The third possible failure of a defrost timer can be breakage of the spacer that separates the contact switches inside the timer, causing a circuit to be made to the compressor and the defrost heater at the same time. If this occurs, the symptoms reported by the customer would be "freezing and thawing." To test for this failure, use a volt/wattmeter as previously described or use an ammeter around the wire leading to the defrost heater, and test for current draw of the defrost heater while the refrigerator is running and trying to cool. You may have to wait a few minutes for this condition to show itself. The defrost termination thermostat would be reacting to the rise and fall of temperature, periodically making and breaking the circuit to the defrost heater.

B.7   DEFROST TERMINATION THERMOSTAT

The *defrost termination thermostat* is a bimetal device that is wired in series with the defrost heater. Positioned in various locations on the evaporator in a frost-free refrigerator, it reacts to the temperature rise resulting from the defrost heater being energized. When the defrost cycle terminates and the run cycle is initiated, the defrost termination thermostat closes as the temperature in the freezer drops.

The defrost termination thermostat usually breaks the circuit after about 10 minutes of heat, allowing the rest of the defrost cycle (usually from 10 to 18 minutes on the defrost timer) to drain water from the evaporator.

A termination thermostat can fail in an open position, which would result in an incomplete circuit to the defrost heater. A frost build-up would occur in the even of failure of the defrost termination thermostat. To test for this condition, proceed as follows:

Step 1: Advance the timer manually to the defrost mode (if it's not already there). Use an ammeter or volt/wattmeter to test for current draw in the defrost mode. If there is no current draw, proceed to Step 2.

Step 2: Unplug the refrigerator, and remove the items in the freezer. Remove the panel covering the evaporator.

Step 3: Locate the defrost termination thermostat and the two wires leading to the component.

Step 4: Cut the two wires, strip them back, and tie them together, eliminating the component from the circuit.

Step 5: Make sure no bare wires are touching any metal surface and plug the refrigerator into the receptacle and test for current draw of the defrost heater. If the defrost heater heats, the termination thermostat must be replaced. An example of the method of mounting the defrost termination thermostat is shown in Figure B-14.

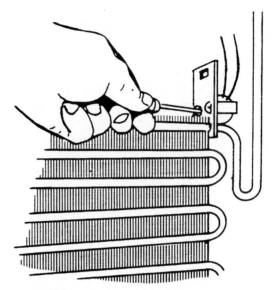

Figure B-14.   Removing a defrost termination thermostat.

It is possible to test a defrost termination thermostat with an ohmmeter. But you will only get a continuity reading if the component has been chilled to a temperature that will allow the bimetal assembly inside to close. Testing a new defrost termination thermostat at room temperature with an ohmmeter will yield an "open" reading.

B.8   DEFROST HEATER

Defrost heaters are found in frost-free refrigerators. They can be in the form of a glass tube containing a heating element or in the form of a metal casing containing a resistive wire. Metal defrost heaters are similar in appearance to the bake element in an oven. In some cases, a defrost heater may be manufactured to "wrap around" an evaporator, and in some models, such as side-by-side refrigerators with a long evaporator, two or three defrost heaters may be used. In the case of multiple defrost heaters, all heaters must be replaced if one fails.

Failure of a defrost heater will result in a frost build-up on the evaporator. To test for the possible failure of the defrost heater, proceed as follows:

Step 1: Disconnect the refrigerator from the wall receptacle.

Step 2: Remove the items from the freezer, and remove the panel covering the evaporator.

Step 3: Locate the leads to the defrost heater, and, with an ohmmeter set on R×100 scale, test for resistance on the component. If the heater does not show continuity or the specified resistance, the heater must be replaced.

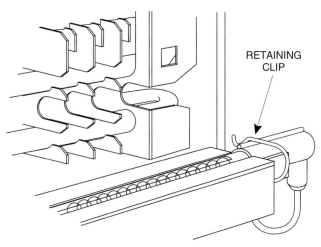

Figure B-15.   One method of positioning a defrost heater.

An alternate test can be to check for voltage applied to the defrost heater. To accomplish this test, disconnect the wires leading to the defrost heater, make sure bared wire ends are not touching a metal surface, and, with the defrost timer in the defrost mode, plug the refrigerator in. If you read 120 volts at the connections to the defrost heater and the heater will not heat when connected, the heater must be replaced. Figure B-15 shows one method of mounting a defrost heater in a trough under the evaporator.

B.9   EVAPORATOR FAN MOTOR

The *evaporator fan motor* is used in a frost-free refrigerator to force the air through the finned evaporator and circulate air throughout the

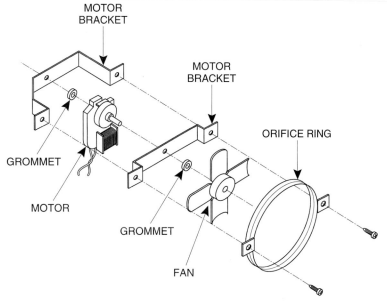

Figure B-16.   An evaporator fan motor circulates air throughout the freezer and fresh food sections.

freezer and fresh food sections of the refrigerator/freezer. A failure of the evaporator fan motor will result in a "running but not cooling" complaint. In some cases, you may find the evaporator fan motor controlled by one or more door switches. To test for operation of the evaporator fan motor, proceed as follows:

Step 1: Disconnect the refrigerator from the wall receptacle.

Step 2: Remove the items from the freezer, and remove the panel covering the evaporator.

Step 3: Make sure the defrost timer is in the run mode and that the temperature control is calling for cooling. Plug the refrigerator into the wall receptacle, and test for 120 volts at the wiring harness that connects to the fan motor. If you read 120 volts at the fan motor and the motor will not run, it must be replaced. One method of mounting an evaporator fan motor is shown in Figure B-16.

B.10 CONDENSER FAN MOTOR

The condenser fan motor in a refrigerator has two purposes:

1. To cool the hot refrigerant vapor as it travels through the condenser
2. To keep the compressor cool

A clockwise (cw) or counterclockwise (ccw) motor may be used depending on the manufacturer's design. Clockwise and counterclockwise motors are rated in watts. A 2 or 2.5 watt condenser fan motor is commonly used.

Failure of the condenser fan motor will cause a "running but no cooling" complaint due to the fact that the refrigerant flowing through the condenser is not being cooled and the compressor may be kicking off on the overload protector. A condenser fan motor that siezes up cannot be lubricated and must be replaced.

Ice Makers, Ice Dispensers, and Chilled Water Dispensers

The automatic ice maker has been a popular accessory in frost free refrigerator/freezers for many years. Fundamentally, there are two distinct styles of ice makers, the flex tray ice maker and the mold style ice maker. In either case, they can be field installed by technicians but in many cases they are installed at the factory. One exception to the field installed ice maker would be the flex tray type that is designed not only to make ice but also to act as a timing device to initiate the defrost cycle. Whirlpool and Sears refrigerators may sometimes be found to use this type of ice maker control system.

In addition to the automatic ice maker, chilled water dispensers consisting of a water tank for storage in the fresh food compartment, a solenoid valve and connecting tubing are added features commonly found on side by side refrigerators and some top mount models.

C.1 MOLD STYLE ICE MAKERS

The mold style ice maker, sometimes referred to as the Whirlpool or Servel style of ice maker is mounted in the freezer section in such a way that it is directly in the air stream from the evaporator fan. In most cases, the wiring necessary to operate the unit is already built into the refrigerator cabinet but with some models a field installable ice maker kit comes complete with the wiring harness that you'll have to install according to manufacturer's instructions. A mold style ice maker is shown in Figure C-1.

The component that works in conjunction with the ice maker is the solenoid valve. Switches inside the ice maker assembly govern

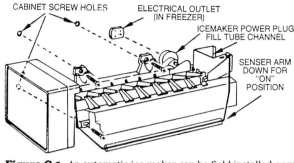

Figure C-1. An automatic ice maker can be field installed or may be installed at the factory.

the operation of the ice maker and also energize the solenoid valve at the appropriate time to allow water to travel through a tube and into the ice maker cavity. An ice maker water valve is shown in Figure C-2.

The mold style ice maker contains three timing switches:

1. A holding switch that assures that once a revolution of the ice ejector has begun, the rotation will be completed
2. A water valve switch that energizes the solenoid of the water valve briefly to allow water to run into the mold
3. A shut off switch that is activated by the sensor arm of the ice maker. This arm, when it settles on ice in a full storage bin will not allow the ice maker to cycle again until some of the ice has been removed.

The holding switch and the water switch are activated by a timing cam with cavities. When the timing cam rotates, the switch buttons are allowed to fall into the cavities (or they ride on the round surface of the cam) and as a result, the switches make and break as necessary. The timing cam is driven by connecting gears that are driven by a low-wattage stall type motor. Figure C-3 shows how the switches are positioned within the ice maker and Figure C-4 shows the motor and gear assembly.

Figure C-2. A solenoid-operated water valve delivers water to the ice maker.

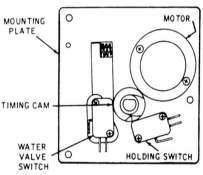

Figure C-3. The three timing switches in a mold style ice maker are the holding switch, water valve switch, and the shut-off switch.

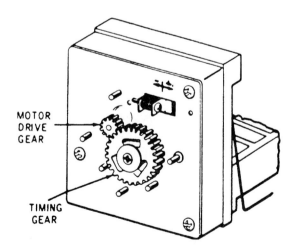

Figure C-4. A motor and gear assembly is used to rotate the ejector in a mold style ice maker.

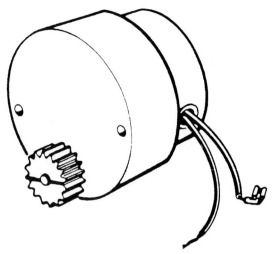

Figure C-5. A mold style ice maker uses a low wattage stall-type motor.

The motor in this type of ice maker is known as a stall type motor because it has to be able to stall while the ice in the mold breaks free. The motor, such as the one shown in Figure C-5, is sometimes similar in appearance to the drive motor on a defrost timer. In many makes and models, the ice maker motor causes the ejector to rotate on full revolution in approximately three minutes.

Two other electrical components are the mold heater and thermostat. The mold heater is a cal-rod style heater and the thermostat is similar in appearance to a defrost termination thermostat. The thermostat initiates the harvest cycle of the ice maker and the mold heater warms the mold to allow the ice (crescent shaped because of the construction of the mold) to break loose and be harvested. Figure C-6 shows a mold heater and Figure C-7 shows the bimetal thermostat.

Depending on the age of the ice maker, the harvest cycle may consist of one or two revolutions. For our purposes, we'll use the two revolution model to explain the cycle of the ice maker electrically

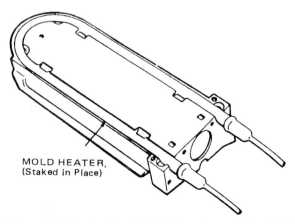

MOLD HEATER,
(Staked in Place)

Figure C-6. A cal-rod type heating element warms the mold to allow the ice crescents to be ejected.

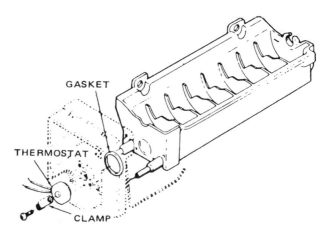

GASKET

THERMOSTAT

CLAMP

Figure C-7. The thermostat that initiates the harvest cycle in a mold style ice maker.

and to illustrate the position of the ejector. Figure C-8 shows the fundamental wiring diagram for the ice maker. Remember that this is illustrated schematically and that the water solenoid is in actuality located near the bottom of the refrigerator cabinet and the ice maker itself is inside the freezer section.

We're beginning our explanation under the assumption that a harvest cycle has just been completed and the ice maker is in the freeze cycle. Water has just been added, and since the mold was warmed during the just-completed harvest cycle, the thermostat has opened and all components are deenergized. Once the water freezers and the mold chills, the harvest cycle begins.

The harvest cycle has begun because the mold has chilled and the bimetal thermostat has closed. The mold heater is now energized and the motor has started to run, resulting in the ejector beginning to move clockwise. Note that the circuit shown (Figure C-9) with the bold line is through the thermostat, then through the NC (normally closed) to C terminal of the holding switch. This circuit only exists for a short time at the beginning of the cycle.

At this point (Figure C-10) the ejector has rotated a few degrees and the holding switch contacts have switched. The circuit to the motor is now through the holding switch only. The reason the system is designed to work this way is to guarantee that a full revolution will occur even if the bimetal warms up and opens. At this point the shut off arm also begins to move upward.

In Figure C-11 the ejector blades have contacted the ice frozen in the mold. The mold heater is still energized in order to help break the ice loose and the motor is stalling until the ice thaws loose from the mold.

Here (Figure C-12) the ejector has begun to eject the ice crescent. The mold heater remains energized. The water switch is closed at this point because of its position near the timing cam, but the water valve solenoid will not be energized because the path of current flow is through the path of least resistance, in this case though the motor and not through the water valve solenoid. At a point later,

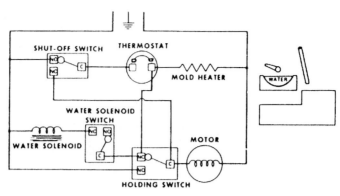

Figure C-8. See text.

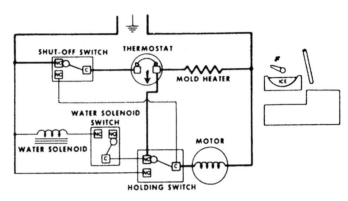

Figure C-9. See text.

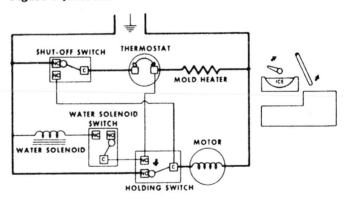

Figure C-10. See text.

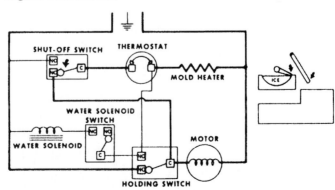

Figure C-11. See text.

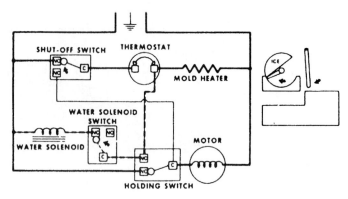

Figure C-12. See text.

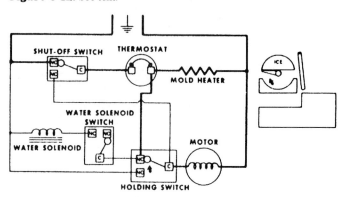

Figure C-13. See text.

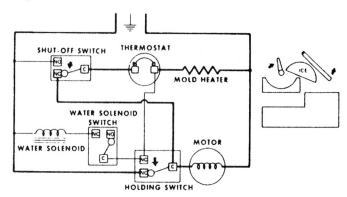

Figure C-14. See text.

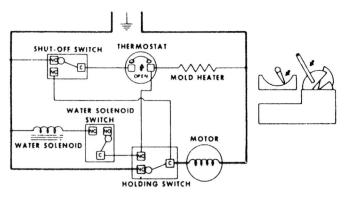

Figure C-15. See text.

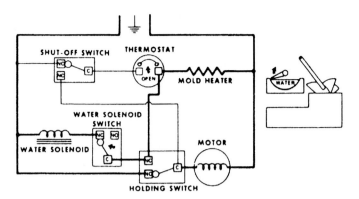

Figure C-16. See text.

when the thermostat has opened, the circuit will complete through the water valve. But for now, since the thermostat is closed and a circuit exists through the mold heater and thermostat, no water will be dumped into the mold.

Figure C-13 shows the end of the first revolution. At this point the heater is still energized and, since the thermostat is still closed, the second revolution begins. As in the beginning of the first revolution, the circuit to the motor is through the closed bimetal thermostat and through the NC and C terminals of the holding switch.

This (Figure C-14) is the second revolution and after the ejector has turned a few degrees, the holding switch once again takes over the circuit to the drive motor. The ice that was harvested is now being dumped into the storage bin. As you can see, the heater is still on and the thermostat is still closed at this point.

In Figure C-15 the heat from the mold heater has now been sufficient to cause the bimetal to open. The heater is deenergized and if the ice bin is not full, the shut-off arm allows the cycle to continue.

Figure C-16 shows that as the ejector nears the end of its second revolution, the water switch closes again and, since this time the thermostat is open and the heater is not energized, the current flow through the water valve solenoid is sufficient to cause the valve to open and allow water flow into the mold. At this point, about 105 of the available volts energizes the water valve solenoid. The remaining voltage of the 120-volt supply drops through the mold heater. With the two revolutions completed, the freeze cycle begins again. When the water freezes sufficiently to chill the mold and close the bimetal thermostat, the harvest cycle will be initiated again.

You can test most mold style ice makers by initiating the cycle manually. Some manufacturers will tell you that it's O.K. to slowly turn the ejector blades clockwise until the holding switch kicks in and takes over the operation of the motor. Some manufacturers, on the other hand, will tell you expressly *not* to turn the ejector by hand. Another method of manually initiating a mold style ice maker cycle is to use a flat screwdriver in the slot of the drive gear, as shown, and turning counterclockwise as shown in Figure C-17.

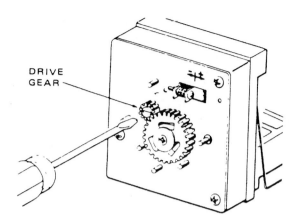

DRIVE
GEAR

Figure C-17. Manually initiating a harvest cycle on a mold style ice maker.

Some mold style ice makers, such as those found in some May-tag refrigerators, do not contain the switches, gears and motor as we've described. In some cases, you'll find a power module is used. While the unit we've described is serviceable and each part can be replaced, ice makers that use a power module cannot be repaired short of replacing the entire power module.

C.2 FLEX TRAY ICE MAKER

The flex tray ice maker differs from the mold style ice maker in that it doesn't use a stationary mold system. Instead, a plastic tray is filled with water. During the harvest cycle, the tray is twisted to force the cubes out. One popular flex tray ice maker is the type found in some Admiral and Magic Chef refrigerators. This type of ice maker, shown in Figure C-18, also uses a bimetal thermostat. In this case though, it's used to initiate the action of a slowly turning timing gear, which allows the unit to operate in freeze time.

Near the end of the freeze cycle, the wire sensor arm drops down into the ice storage bin and the harvest cycle is initiated. Figure C-19 shows the tray at the beginning of the harvest cycle. In Figure C-20, the tray is shown as it engages a stop which causes the tray to warp and twist. As the stop is retracted, the tray snaps, shucking the ice cubes into the storage bin located under the ice maker.

As the cycle continues, the ice maker tray continues to turn and when it is near its upright position, water is allowed to flow into the tray when the solenoid valve is energized. Figure C-21 shows the end of the harvest cycle of this particular flex tray ice maker.

In the event of a component failure within this unit, the entire ice maker head is replaced. Switches, gears and the motor are not replaced individually. In some cases, it is necessary to replace the plastic tray due to deposits on the plastic surface that prevent the ice cubes from falling free when the tray is twisted.

There are several different models of flex tray ice makers used in

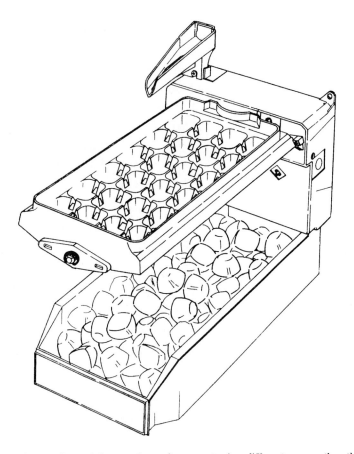

Figure C-18. A flex tray ice maker operates in a different manner than the mold style ice maker.

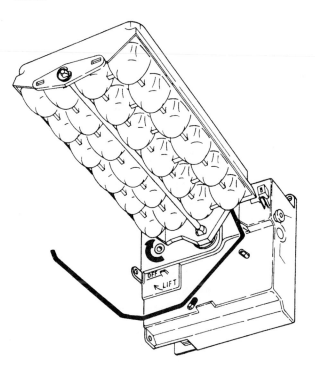

Figure C-19. The flex tray ice maker at the beginning of a harvest cycle.

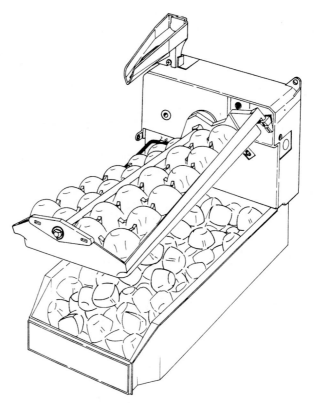

Figure C-20. A flex tray ice maker shucks the cubes when the tray becomes twisted, then snaps loose in the harvest cycle.

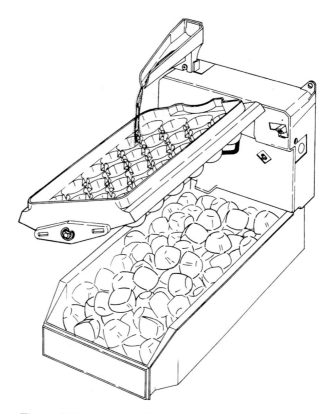

Figure C-21. At the end of a harvest cycle, the flex tray is refilled with water.

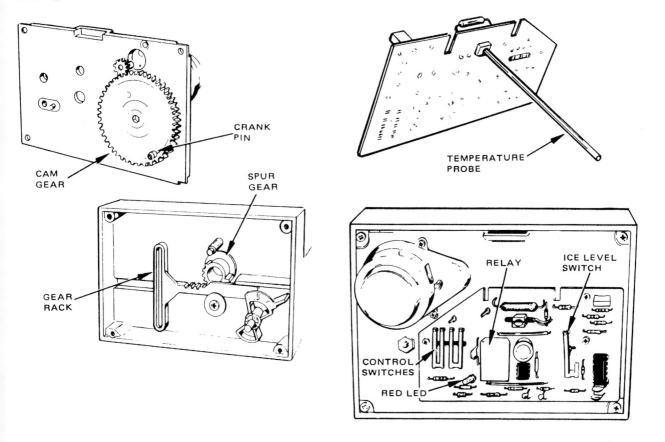

Figure C-22. Some flex tray ice makers are constructed to allow disassembly and component replacement.

refrigerators. Some are of a solid-state design in regard to electrical components and can be serviced with separate parts, whether electrical or mechanical, being replaced. Figure C-22 shows a different type of flex tray ice maker that can be disassembled and serviced in the event of a motor, circuit board, or gear failure.

C.3 CHILLED WATER SYSTEM

Chilled water systems are commonly found in side by side refrigerators. In a few cases, you may see a top mount refrigerator that has an ice maker and a through-the-door chilled water dispenser system. When a unit has an ice maker as well as a chilled water system, a dual solenoid valve is used. One side of the valve supplies water to the ice maker during the cycle, while the other section of the valve supplies water on demand when energized.

In a refrigerator equipped with a chilled water system, the water reservoir is located in the fresh food compartment. Plastic tubing is routed under the refrigerator cabinet and up to a dispensing nozzle. When a glass is pressed against a lever, a switch behind the lever completes the circuit to the solenoid valve and water flows

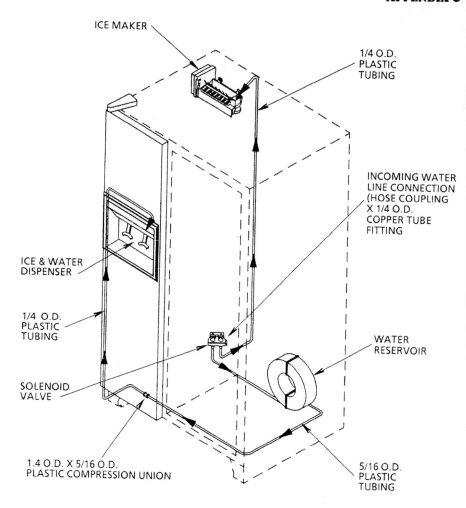

ICE MAKER

1/4 O.D.
PLASTIC
TUBING

INCOMING WATER
LINE CONNECTION
(HOSE COUPLING
X 1/4 O.D.
COPPER TUBE
FITTING

ICE & WATER
DISPENSER

1/4 O.D.
PLASTIC
TUBING

WATER
RESERVOIR

SOLENOID
VALVE

1.4 O.D. X 5/16 O.D.
PLASTIC COMPRESSION UNION

5/16 O.D.
PLASTIC
TUBING

Figure C-23. A chilled water system in a side by side refrigerator. The reservoir is located in the fresh food compartment.

from the nozzle. One example of a chilled water system is shown in Figure C-23.

Water reservoir types vary. The system may use a rectangular, or cylindrical, plastic tank located behind a panel at the rear of the food compartment. In some cases the water reservoir may not even be a tank at all. Instead, it may be a coil or bundle of tubing that can contain a given amount of water in volume.

Refrigerator/Freezer Cabinet Servicing

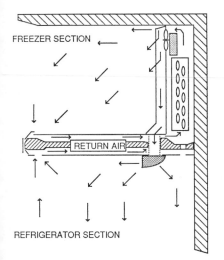

FREEZER SECTION

RETURN AIR

REFRIGERATOR SECTION

Figure D-1. Urethane foam insulation (represented by the diagonal lines) is used to prevent heat migration into the refrigerator cabinet.

D.1 REFRIGERATOR INSULATION

Most refrigerators manufactured today utilize urethane foam cabinet insulation to prevent heat gain in the freezer and fresh food sections. The insulation in the door panels is traditionally a fiberglass material, although there may be exceptions. Using a fiberglass instead of foam insulation in the door assemblies allows for easier replacement of the inner door liners, should that become necessary. Styrofoam is also used, usually as a divider between the fresh food and freezer compartments of a top mount refrigerator/freezer. Figure D-1 shows a side view of a typical refrigerator/freezer cabinet that uses a foam insulation between the inner liner and outer cabinet assembly.

One thing you should remember about servicing a unit with a urethane blown-foam cabinet insulation is that the propellant used to accomplish the insulation process is a CFC, containing the same chemicals used in the refrigerant found in the refrigerator sealed system. When using an electronic leak detector to find a refrigerant leak in this type of unit, the leak detector can be "fooled" into reacting, especially if you're tracing a possible leak on tubing that runs up into the insulated cabinet.

The styrofoam insulation, such as that found as a divider between refrigerator cabinet sections (show in Figure D-2), does not cause the leak detector to render a false reading.

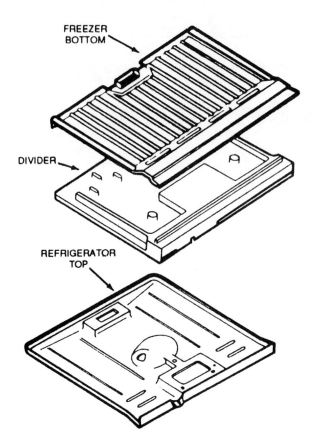

Figure D-2. A styrofoam insulator is often used as a separator between the fresh food and the freezer sections of a top mount refrigerator/freezer.

D.2 DOOR GASKETS

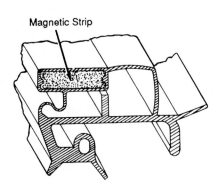

Figure D-3. A magnetic-type door gasket.

On older, latch door refrigerators, a vinyl rubber type gasket was used to provide a good seal between the door and the cabinet. When the latch assembly was properly adjusted, the gasket compressed slightly when the door was closed and the gasket contacted the cabinet. One common door seal test used on these models of refrigerators was the "dollar bill" test. If, when the door was closed, the pressure was enough to cause an inserted dollar bill to tear, the seal was deemed sufficient. This same test performed on newer units that utilize magnetic type door gaskets is not an effective evaluation of the door seal.

Magnetic door gaskets are of vinyl construction with a magnetized strip that fully contacts the entire perimeter of the refrigerator cabinet on some models, and on three sides of the door assembly (with the exception of the hinge side) on others. A section of a magnetic door gasket is shown in Figure D-3.

In some cases, manufacturers secure the gasket to the door with a metal strip that cinches down on the rounded section of the gasket. When this is the case, the screws holding the metal strip do not have to be removed entirely to facilitate removal of the gasket. Loosening

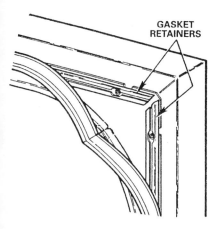

Figure D-4. Metal strips serve to fasten the door gasket as well as to maintain alignment of the door and secure the inner door liner.

them is sufficient. Other manufacturers use a gasket that does not use the rounded section. In those cases, securing screws pass through a hole in the bottom flap of the gasket. Replacing a gasket of this type requires removal of the screws which will also result in removal of the door's inner liner.

Regardless of which type you're servicing, the screws that secure the gasket also serve to keep the door assembly from "racking" or becoming crooked and not aligning with the cabinet. Loosening the screws, aligning the door to match the cabinet, then retightening the screws is the method used to bring the door back into alignment with the cabinet and assuring that the gasket seals properly. Figure D-4 shows how metal strips are used as a retainer to secure the gasket and keep the inner door panel and outer door panel aligned.

D.3 SHELF SUPPORTS

Figure D-5. A twist-type shelf support is both removed and installed by twisting it clockwise.

Shelf supports become damaged due to customer misuse, either when a shelf has been leaned or pulled on, or when an item that won't fit properly between two shelves has been forced into the refrigerator. Due to the fact that some refrigerators use a porcelain-on-steel liner while others may use liner made of a plastic-based material such as polypropylene, different types of shelf supports are used.

With a porcelain-on-steel liner, a twist type support, such as the one shown in Figure D-5, is used. A locking wire on the support cinches it into the hole in the cabinet. This type of support is removed and installed by twisting it clockwise. Remember, when installing this type of support, to insert it 180° off the final intended position. Twisting it will then lock it into the intended position.

On refrigerators that use a plastic liner, other types of shelf supports will be used. They may use a screw and grommet type of mount or a pin that is driven in to fasten the support once it has been inserted into the hole in the liner. Figure D-6 shows these two types of supports.

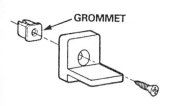

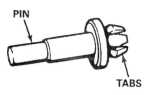

Figure D-6. Other types of shelf supports used in refrigerators with plastic liners include the screw and grommet support and the drive-pin type of support.

Installation Considerations for Major Appliances

E.1 DISHWASHERS

The dishwasher changeout is the most common appliance installation you'll have to accomplish as an appliance technician. The term changeout means that an undercounter dishwasher is already in place and cabinet work will not be necessary. The three things to be considered in a changeout are the water supply line, drain line, and electrical supply.

The minimum requirement for a water supply line is ³/₈ inch, in most cases a soft copper rather than hard drawn copper is used. The soft copper tubing allows for some flexibility in rerouting the water supply line in the event that the water valve on the replacement dishwasher is in a different location (on the right instead of the left side of the machine, for example) or in a slightly different position.

In cases where the dishwasher was installed at the time the home was built, the water line may be routed through the floor or may come directly out of the wall. In cases where the dishwasher was added later, the water line will most likely be teed off the hot water line under the kitchen sink. Figure E-1 shows the three methods of routing water supply lines.

In either case, a separate shutoff valve should be used. In the event that a water inlet valve sticks in the open position or must be replaced due to solenoid coil failure, a separate shutoff valve on the dishwasher supply line will eliminate the necessity to shut down an entire area or the main water supply.

The dishwasher water inlet valve itself accepts a ½" male pipe thread fitting. The final connection could be a sweat fitting, a flared connection, or a compression fitting that uses a nut and ferrule.

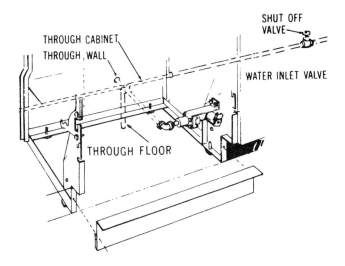

Figure E-1. The water supply line for a dishwasher can be routed through the floor, from the wall or through cabinets from the water supply under the sink.

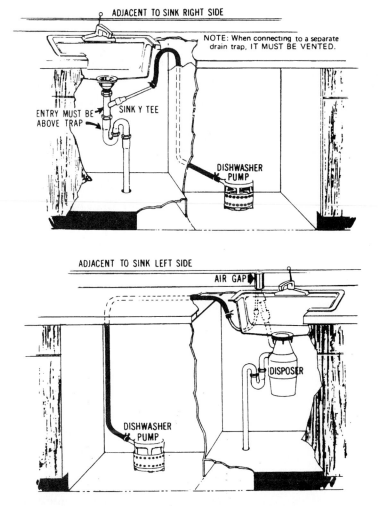

Figure E-2. A dishwasher drain hose must be installed to prevent siphoning from the drain system or garbage disposer.

Caution! *In the event that the equipment you're working with has a sweat fitting and you've decided to stay with that method, use extreme caution when working with a torch. Excessive heat can damage the dishwasher water valve.*

The dishwasher drain line is, in most cases, routed from the pump assembly of the dishwasher to the drain line under the sink. In the event that a garbage disposer is used, the dishwasher drain hose can be connected to the stub provided on the disposer. When this method of connection is used, a knockout plug inside the disposer must be removed. Figure E-2 shows two examples of a dishwasher drain line.

An important consideration for the dishwasher drain line is finding a method of providing a siphon break so that dirty water from the sink or disposer cannot run down into the dishwasher pump assembly. When routing a dishwasher drain hose, be sure that a section of the hose is secured at a location above the water level in the sink or that an air gap system is used.

A dishwasher requires a 15-amp circuit independent of other kitchen appliances. One exception to this rule may be the garbage disposer. You may find a duplex receptacle under the sink into which both the dishwasher and disposer are plugged in. The disposer may operate on a wall switch, meaning that the receptacle is wired as a split receptacle, with one half "hot" all the time and the other half of the receptacle hot only when the switch on the wall is flipped to the on position. In some cases, the dishwasher may be hard wired, meaning that the power supply does not come from a wall receptacle, but instead appears as romex wired directly to the dishwasher.

In either case, the hot and neutral wires must be properly connected to the dishwasher lead wires and the ground wire must be properly installed as shown in Figure E-3.

Ground Screw

MACHINE MUST BE PROPERLY GROUNDED

Figure E-3. A dishwasher must be properly grounded for consumer safety.

E.2 REFRIGERATORS

Electrical supply for a refrigerator is usually a separate 20 amp circuit. As discussed in previous chapters, the safety grounded 3-prong plug is the type of power cord used on a refrigerator. In the event that a 3-prong receptacle is not available, the grounding prong on the plug is not to be cut off. Instead, the electrical supply should be modified to bring it up to standard, providing for a safety ground system. In some cases, you may find that an adapter has been used to allow the 3-prong plug to connect to a 2-prong receptacle. More often than not, the use of the adapter is not accomplished in such a way as to allow for the appliance to be grounded.

Ventilation around the cabinet is another consideration in the installation of a refrigerator. In the event the unit is equipped with a static condenser, 2 inch clearance from the wall should be allowed and 4 inch should be allowed from the top of the refrigerator to any

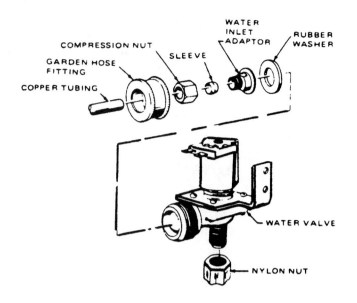

Figure E-4. A typical ice maker water valve.

kitchen cabinets. This clearance is necessary to allow for the dissipation of heat from the condenser tubing mounted on the back of the refrigerator cabinet. A unit equipped with a fan cooled condenser does not require clearance for ventilation since the fan draws air in from the front of the cabinet and discharges it out the opposite side of the cabinet front.

You may find yourself in a situation where you have to run a water supply line to a refrigerator. In newer homes, a shut-off valve will be located in a small recessed area directly behind the refrigerator. In most cases, this valve will accept the ¼" water line. In a situation where there is no shut-off valve, you'll have to route a water line from the refrigerator to a water source, usually the cold water supply line to the kitchen sink or through the floor to a water line in the basement.

When working with copper plumbing, a self-piercing valve is commonly used to tap into the water line. Piercing valves may also be used on galvanized plumbing, but drilling a hole in the pipe is necessary.

Be sure to leave a coil of tubing behind the refrigerator. This allows for the movement of the refrigerator without kinking the water line. The final fitting to use when connecting to the refrigerator ice maker water valve will usually be a ¼ inch compression union. The ice maker water valve (or dual solenoid water valve in the case of a unit equipped with a chilled water feature) is equipped to accept a garden hose fitting that adapts to ¼ inch copper tubing. Refer to Figure E-4.

Many models of refrigerators are designed to allow for a change in door swing. It may be inconvenient because of kitchen design for a door to open in a particular direction, making the modification nec-

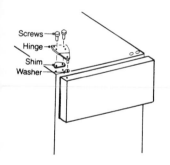

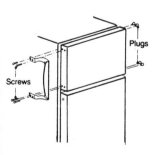

Figure E-5. Reversing door swing requires repositioning and relocation of the door hinges and handles.

essary. Changing the door swing requires removal of both doors in the case of a top mount or bottom freezer model, and relocation of the hinges and repositioning of the door handles. The threaded openings required to accomplish the repositioning and relocation are already part of the cabinet assembly and are filled with decorative plastic plugs as shown in Figure E-5.

E.3 AUTOMATIC WASHERS

When installing an automatic washer, always check to make sure the water supply lines are properly connected and that the drain line is secured to prevent it from popping out of the standpipe or the sink. In the event that hot and cold water hoses are reversed, the machine will not operate properly in cycles with specific water temperature features. Figure E-6 shows the installation of automatic washer supply and drain hoses.

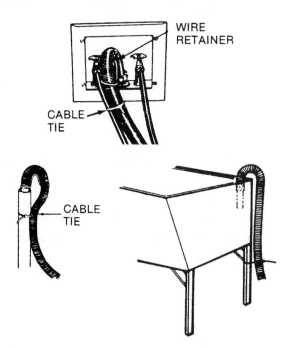

Figure E-6. Proper method of installing automatic washer water supply hoses and drain hose.

The drain hose from the washing machine should be positioned in a standpipe of the sink above the level of the water in the washer tube. Leaving the hose too low can result in siphoning, which prevents the washer from filling up in the rinse cycle. In the event that you need to accomplish an installation to a floor drain or similar situation, a siphon break assembly can be installed on the washer to prevent siphoning. It's also important that the drain hose fits loosely in the standpipe. A hose that fits too tightly can also cause siphoning.

It's important that an automatic washer be level and firm. Some machines are equipped with self levelers on the rear of the unit, allowing you to level the machine by raising the rear of the unit a reasonable distance and dropping it onto the floor. A level may be used to check the machine for proper position. Test for level by making sure that the machine does not wobble when applying pressure corner to corner.

If the floor is not firm, the machine will walk and will become unlevel in the spin cycle. In severe situations it may be necessary to set the machine on a section of $\frac{3}{4}$ inch plywood to prevent shimmy on a floor that is not firm enough to support the machine in the spin cycle.

E.4 CLOTHES DRYERS

Improper venting can seriously affect the operation of a clothes dryer. Most manufacturers don't recommend flexible material in venting a dryer due to the fact that it allows for a build-up of lint, affecting the ability of the dryer to heat properly. Figure E-7 shows one manufacturer's comparison of a correct and incorrect venting installation.

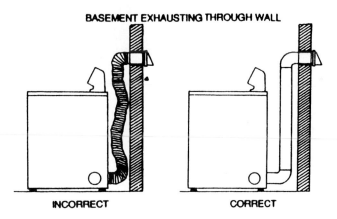

Figure E-7. Rigid venting material should be used to properly vent a dryer.

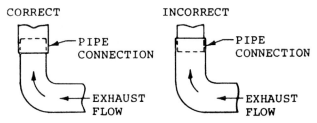

Figure E-8. Improperly connected piping joints can create lint build-up in the venting system.

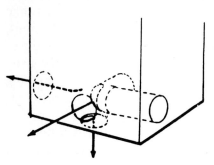

Figure E-9. A dryer can be modified to accomplish venting.

Another factor to consider in the installation of venting material is properly connected joints. As Figure E-8 shows, improper installation of piping connectors can result in a build-up of ling.

In most cases, the dryer vent is positioned out of the rear of the dryer cabinet, however, most dryers are equipped to allow the modification of the vent piping to exit through either side of the cabinet or the bottom of the dryer if necessary. Figure E-9 shows the method of modifying a dryer vent system.

Index